# COMPOSTING and RECYCLING

## MUNICIPAL SOLID WASTE

# COMPOSTING and RECYCLING
## MUNICIPAL SOLID WASTE

By

Luis F. Diaz
George M. Savage
Linda L. Eggerth
Clarence G. Golueke

CalRecovery, Inc.
Hercules, California

LEWIS PUBLISHERS
Boca Raton Ann Arbor London Tokyo

**Library of Congress Cataloging-in-Publication Data**

Composting and recycling municipal solid waste / by L. F. Diaz ... [et al.].
p. cm.
Includes bibliographical references and index.
ISBN 0-87371-563-2
1. Refuse and refuse disposal—Biodegradation. 2. Recycling (Waste, etc.) I. Diaz, Luis F.
TD796.5.C59 1993
628.4'458—dc20 92-36987
CIP

Direct all inquiries to CRC Press, Inc., 2000 Corporate Blvd., N.W., Boca Raton, Florida 33431.

PRINTED IN THE UNITED STATES OF AMERICA
1 2 3 4 5 6 7 8 9 0
Printed on acid-free paper

# The Authors

## Luis F. Diaz

Luis F. Diaz received his Doctorate in Environmental Engineering from the University of California at Berkeley. He has been involved in the field of waste management for over 19 years. Dr. Diaz has conducted numerous waste management studies, technical and economic assessments of refuse processing systems, environmental analyses of resource recovery systems, and marketing studies for materials and energy recovered from residues. He has also carried out a large number of projects involving the biological conversion of organic matter into useful products. Dr. Diaz has conducted technical, economic, environmental, and institutional assessments of solid waste management systems for state and federal agencies, private industries, and international lending institutions, as well as governments in Asia, Africa, Europe, and Latin America. Dr. Diaz has authored or coauthored seven books and more than 100 publications in the fields of waste and energy management.

# George M. Savage

George M. Savage received his M.S. in Mechanical Engineering from the University of California at Berkeley. He has more than 20 years of experience in the field of solid waste management. Mr. Savage has participated in a wide variety of projects, including technical and economic feasibility studies of waste management alternatives, design and construction of processing facilities, field testing of collection and processing systems, development of test methods and performance criteria in the field of solid waste management, waste characterization studies, and market development projects. He often serves as an expert consultant to municipalities and private industry in matters pertaining to the design, procurement, construction, operation, and performance of solid waste handling, processing, and conversion systems. He is a registered professional engineer in several states. Mr. Savage is the author and co-author of several books and more than 150 publications in the field of solid waste management.

## Linda L. Eggerth

Linda L. Eggerth received her degree in Business Management from St. Mary's College in Moraga, California. She has been involved in the field of waste management for over 15 years. Ms. Eggerth has both directed and participated in a large variety of waste management and resource recovery projects and has developed an international expertise in several areas including preparation of request for proposals, financial and economic analysis of waste collection and processing alternatives, marketing of recovered materials, and the development of guidelines for the classification of compost. She has played key roles in assisting communities in the implementation of recycling and composting programs, and has participated in international projects involving material and energy recovery from solid wastes. She has authored or co-authored several articles and technical reports in the field of solid waste management and is regularly requested to make presentations at seminars and conferences.

## Clarence G. Golueke

Clarence G. Golueke received his Doctorate in Botany from the University of California at Berkeley. Dr. Golueke has over 40 years of experience in the fields of solid and liquid waste management. He is an internationally recognized authority in the biological conversion of solid wastes. Since the early 1950s, he has been continually involved in all aspects of composting and anaerobic digestion of municipal solid wastes, agricultural residues, industrial wastes, and various types of sludges. Dr. Golueke was one of the first scientists to delineate the basic parameters for the application of composting in the field of solid waste management. He has provided expert advice to national and international organizations in all aspects of solid waste management. Dr. Golueke is a member of editorial boards of several journals dealing with waste management and environmental control. He has published several books and more than 200 articles and technical reports in the fields of solid and liquid waste management.

# Table of Contents

# Composting and Recycling Municipal Solid Waste

# CHAPTER 1

# Introduction

## DEFINITION AND CLASSIFICATION OF MUNICIPAL SOLID WASTE

The first term to be defined is ''waste'' as it applies to municipal waste. In response to the increasing emphasis on recycling, the term ''waste'' as used in this book applies to ''a resource discarded by its possessor or user (dweller, commerce, industry, government) because apparently it is of no further use to the possessor.'' Municipal wastes can be classified on a number of bases, but for this book the class of municipal waste that is of interest is solid waste. Definitions of municipal solid waste (MSW) encountered in the literature range in degree of complexity. At one extreme are those favored by regulatory bodies and which characteristically are verbose versions couched in confusing legalese and so all-inclusive as to have little practical utility. At the other extreme are the realistic and practical versions favored by solid waste management practitioners. A logical ''definition'' of MSW is ''a solid waste generated by a community (municipality).'' As with municipal wastes in general, MSW can be and is classified on several bases. The definitions and classification currently in use are summarized in Table 1.1. A recent trend is the treatment of yard waste (garden debris, prunings, and leaves) as a distinct waste stream.

## NATURE OF MSW

Regardless of the developmental status (industrialized vs developing), MSW generated in large, densely populated metropolitan areas, particularly the ''inner city'' sections, differs somewhat from that generated in the suburbs and considerably from that generated in small communities and rural areas. Primarily, the difference takes the form of change in concentration of organic matter as well as a change in concentration of paper and paper products. The changes are more pronounced in developing countries and in those that are not highly industrialized. The changes are illustrated by the data in Table 1.2.

**Table 1.1 Municipal Solid Waste Materials by Kind, Composition, and Sources[a]**

| Kind | Components | Primary Sources |
|---|---|---|
| Garbage (food wastes) | Wastes from preparation, cooking, and serving of food; market wastes; wastes from handling, storage, and sale of produce | Households, restaurants, grocery stores, markets |
| Rubbish | Combustible: paper, cartons, boxes, barrels, wood, excelsior, wood furniture, bedding, dunnage, plastic | Households, commercial businesses, industry |
| Yard wastes | Leaves, garden debris, trimmings, pruning | Residences, public grounds |
| Street sweepings | Sweepings, dirt, leaves, catch basin dirt, contents of litter receptacles | Municipalities |
| Ashes | Residue from fires used for cooking and heating and from on-site incineration | Households, incinerators, industry |
| Hazardous | Toxic, pathogenic, highly flammable, explosive radioactive materials | Households, hotels, hospitals, institutions, stores, industry |
| Dead animals | Cats, dogs, horses, cows | Streets, sidewalks, alleys, vacant lots |
| Abandoned vehicles | Unwanted cars and trucks left on public property | |
| Industrial wastes | Food processing wastes, boiler house cinders, lumber scraps, metal scraps, shavings | Factories, power plants |
| Demolition wastes | Lumber, pipes, brick, masonry, and other construction materials from razed buildings and other structures | Demolition sites to be used for new buildings, renewal projects, expressways |
| Construction wastes | Scrap lumber, pipe, other construction materials | New construction, remodeling |

[a] Adapted from Reference 5.

## IMPORTANCE OF A SOUND SOLID WASTE MANAGEMENT PROGRAM

The satisfactory solid waste management requirement is applicable at all stages in the handling of municipal solid wastes, i.e., beginning with the generation of the waste and continuing through storage at the generation site, through collection and transport, and eventually on to treatment and then ultimate disposal. The rationale underlying the requirement is that a good quality of the environment is a necessity — not a luxury.

The significance of the requirement is proportional to the severity of the penalty invariably incurred by failure to design and implement a proper management program. An early failure or neglect is followed by a severe penalty at a later date in the form of a legacy of a hopeless deterioration or even needless loss of resources and a staggeringly adverse impact is exerted on the environment and on the physical and mental well-being of the affected inhabitants. The severe penalty for neglect prevails regardless of the developmental status of a country.

**Table 1.2 Quantity and Composition of Municipal Solid Waste[a] (Percent Wet Weight)**

| Material | India, urban[1] | Manila, Philippines[3] | Asuncion, Paraguay | Lima, Peru | Mexico City, Mexico[2] |
|---|---|---|---|---|---|
| Putrescibles | 75.0 | 48.8[b] | 60.8[b] | 34.3 | 56.4[b] |
| Paper | 2.0 | 17.0 | 12.2 | 24.3 | 16.7 |
| Metals | 0.1 | 1.5 | 2.3 | 3.4 | 5.7 |
| Glass | 0.2 | 5.3 | 4.6 | 1.7 | 3.7 |
| Plastics, rubber, leather | 1.0 | 6.5 | 4.4 | 2.9 | 5.8 |
| Textiles | 3.0 | 3.7 | 2.5 | 1.7 | 6.0 |
| Ceramics, dust, stones | 19.0 | 17.2 | 13.2 | 31.7 | 5.7 |
| Weight/capita/day (lb) | 0.91 | 0.88 | 1.41 | 2.12 | 1.50 |

| Material | Caracas, Venezuela | Sasha Settlement Ibadan, Nigeria | Berkeley, CA U.S. | Broward Co. Florida U.S. |
|---|---|---|---|---|
| Putrescibles | 40.4[b] | 76.0 | 39.0[c] | 39.8[c] |
| Paper | 34.9 | 6.6 | 40.1 | 37.8 |
| Metals | 6.0 | 2.5 | 3.0 | 5.6 |
| Glass | 6.6 | 0.6 | 7.6 | 6.7 |
| Plastics, rubber, leather | 7.8 | 4.0 | 6.3 | 9.0 |
| Textiles | 2.0 | 1.4 | 1.7 | — |
| Ceramics, dust, stones | 2.3 | 8.9 | 2.3 | 1.1 |
| Weight/capita/day (lb) | 2.07 | 0.37 | 3.10 | 3.86 |

[a] Based on actual measurements.
[b] Includes small amounts of wood.
[c] Includes yard wastes, food wastes, and wood.

Municipal waste usually contains many materials which, if retrieved and suitably processed, would constitute resources that can be of great utility.

The putrescible organic fraction of MSW exerts an adverse impact by way of generating foul odors and of attracting and serving as food and shelter for vectors of microorganisms pathogenic to humans. Obnoxious and possibly toxic decomposition products from organic wastes contaminate the air, water, and land resources. MSW also contains many toxic nonputrescible organic and inorganic substances which, if not properly managed, also eventually contaminate the air, water, and land resources.

The penalties mentioned in the preceding paragraph can not be avoided by a resolve to do something about the wastes in the future. On the contrary, postponement aggravates the penalties. Unfortunately, nature is not affected by rationalizations. The more the quality of the environment is degraded, the greater is the effort required to restore its good quality. Nor is the increase in need in simple proportion (1:1) to degree of degradation, rather it is more like a geometrical progression.

Finally, for a developing nation, the attainment of full development (industrialization) is of no avail if the quality of the environment has in the meantime become unbearable. Ideally, development is sought as a means of improving the lot of the citizenry. Because a livable environment is essential to the well-being of the citizenry, the care given to the preservation of a good environment should be commensurate with that afforded to the attainment of full development.

## RESOURCE RECOVERY ASPECTS

The resource recovery aspects of the organic component are threefold. Technology is available for transforming it into compost (soil amendment), for converting it into methane (combustible gas), and for hydrolyzing the cellulosic components into glucose. The glucose can serve as a substrate either for single-cell protein production or for fermentation into ethanol. Of the three conversions, compost has been and is the most frequently applied. Many hurdles, primarily economic in nature, must be surmounted before either single-cell protein production or ethanol production become a practical reality, although the outlook is propitious for biogasification.

Despite being commonly used in wastewater treatment for processing settleable sewage solids, until relatively recently, methane production (''biogasification'') has been rather infrequently used for the treatment of solid wastes and extremely seldomly for treating MSW. Since the early 1980s, it has begun to receive a slowly increasing amount of attention for its waste treatment potential and as a possible source of energy. The upswing in attention is partly a phenomenon of a growing worry about the possibility of a future energy shortage. However, an influence on the developing interest not to be overlooked is the favorable impression made by the successful tapping of the methane formed in landfilled MSW.

In addition to the three biological conversions, another application is the use of organic wastes as a feedstock in chemical and thermal processes in which combustible gases and oils are produced. Finally, they can be and are used directly as a fuel in combustion operations designed to produce heat and generate steam.[4]

The detailed explanation and description of the methods outlined in the preceding paragraphs for exploiting the organic wastes are given in later chapters. The recovery of inorganic resources (e.g., ferrous and aluminum recovery) is described in terms of segregation at the point of generation. Additional mechanical processing of these materials at a central facility also is discussed.

## SCOPE OF THE BOOK

Although this book is intended to serve as a reference book, it could be classified as being a textbook were it not for the absence of problem sets and assignments. Its objective is to identify, describe, explain, and evaluate the options available when composting and recycling MSW are key goals of waste management. It is hoped that the book will be useful and interesting not only to those who are directly responsible for solid waste management, but also to those who otherwise have a role in it and to individuals interested in or concerned with MSW composting and recycling.

In line with these goals, the book supplies the background and basic information needed to arrive at rational decisions, i.e., those in keeping with available economic and technological resources.

The book is organized into ten chapters. The first chapter is an introductory chapter in which the nature of MSW is discussed in terms of definitions, classification, and some discussion on composition. Not surprisingly, it emphasizes the importance of having a sound solid waste management program and refers to aspects of resource recovery. It closes with this section on the scope of the book. Chapter 2 deals with storage and collection as elements of MSW management in general and as geared to recycling, particularly biological recycling. It has a special section on the collection of recyclables. In Chapter 3 are stressed and explained the need and procedures involved in determining the quantity, composition, and key physical characteristics of the MSW to be managed and processed. Chapter 4 classifies and describes the steps required for processing MSW for resource recovery. In Chapter 5, these processes are applied to recyclables, particularly the design and operation of materials recovery facilities (MRFs). Chapter 6 deals with the use of uncomposted organic matter as a soil amendment. Composting in all of its aspects is discussed in Chapter 7. Thus, the chapter deals with various aspects of composting and discusses the basic principles and technology of the process and the use of the compost product. The marketing of the important types of recyclables and compost is described and discussed in Chapter 8. Chapter 9 gives an overview of biogasification, a biological reclamation process that is beginning to attract attention as having a potential in MSW management. Chapter 10 explores integrated waste management, a movement that is gaining momentum. The material covered in the nine chapters preceding Chapter 10 is of especial value in integrating waste management.

## REFERENCES

1. Flintoff, F., "Management of Solid Waste in Developing Countries," WHO Regional Publications, South-East Asia Series No. 1, New Delhi (1976).
2. Savage, G.M., "Solid Waste Management in Mexico City, Mexico," Report to the Pan American Health Organization, February (1979).
3. Diaz, L.F., "Solid Waste Management in Metropolitan Manila," Report to the World Bank, December (1978).
4. Diaz, L.F. *et al.*, "Market Potential of Materials and Energy Recovered from Bay Area Solid Wastes," College of Engineering, University of California, Berkeley, March (1978).
5. *Municipal Refuse Disposal,* 3rd ed. (Chicago: Public Administration Service, Institute for Solid Waste of American Public Works Association, 1970).

# CHAPTER 2

# Storage and Collection

## INTRODUCTION

The successful conduct of a resource recovery undertaking is strongly influenced by the storage, collection, and transport steps in municipal solid waste (MSW) management in that the three steps can either simplify or complicate the undertaking. Heretofore in general practice, the effect has been one of complication due to an indiscriminate intermixing of recyclable and nonrecyclable resources. The mixing begins with storage and is compounded in collection and transport.[1,2] Because of the spread of the practice of source separation, this unfortunate occurrence is now being avoided in an increasing number of undertakings.

Much of the information provided in this and in succeeding chapters is not restricted to activities directly geared to recycling and composting. On the contrary, it is expanded to apply to MSW management as a whole. This expansion is made to provide a framework and background for the more specific information regarding recycling and composting.

## STORAGE AT THE POINT OF ORIGIN

MSW must be stored at the point of origin because of the discontinuous nature of solid waste collection. At this time, conditions characteristic of almost all communities render pipe transport the only feasible means of removing a waste on a strictly continuous basis.[3,4] However, many problems combine to make pipe transport of MSW impractical. Chief among them is the fact that pipe transport places a premium upon homogeneity, imposes an upper limit on particle size and density, and demands some degree of fluidity. On the other hand, an outstanding characteristic of MSW is its heterogeneity. Moreover, it includes a sizeable fraction that has particle sizes and densities greater than those accommodated by pipe transport. MSW is almost completely devoid of fluidity. Consequently, unless subjected to considerable processing, MSW does not lend itself to pipe transport. Therefore, it seems that for the time being, continuous removal of MSW from the point of generation is impractical, and hence, the only recourse is to store the waste at its point of origin until it can be transported to the site of disposal or processing.

## Principles

One of the basic requirements for proper storage of MSW is that the container or containers in which the waste is stored be designed and constructed such that it ensures isolation of the waste from the environment. From this requirement it follows that a prime measure of satisfactory storage is the degree of isolation. The extent or degree of isolation required is largely a function of the gravity and intensity of the adverse effect that potentially would be imposed on public health and quality of the environment, if the waste were not isolated or only partially isolated. Thus, the amount of isolation to be provided for a highly putrescible waste would be far greater than that imposed upon a relatively stable material. Much of the waste destined for composting would be in the putrescible category, and hence must be more completely isolated than would be needed with the nonputrescible categories.

Additional requirements for storage at the point of origin include the following: (1) In situations in which space is relatively valuable, storage of the waste should involve a minimum amount of space. The reason is that space occupied by wastes is nonproductive, and hence in a commercial setting would represent a loss of income. In a residential setting, it ties up part of the very limited amount of space available to the average resident. (2) The container together with the waste in it should be readily accessible for collection. This requirement ensures efficient use of manpower and tends to minimize spillage. (3) The cost involved should be the lowest that is commensurate with satisfactory storage.

### *Containers*

Aside from those mentioned in the preceding paragraphs, specific requirements for containers are (1) it can be closed with a tight-fitting lid; (2) it is fabricated of a durable material (metal or rigid plastic); and (3) its capacity should be such that it can be safely manipulated by the collection crew.

Since multimaterial collection has increased in popularity, various types and sizes of storage containers currently are manufactured. Typical range of sizes fluctuate from about 5 to 90 gal for residential use. Some of the containers are manufactured using up to 100% recycled plastic. Samples of various types and sizes of containers are shown in Figure 2.1.

Innovative developments in the design of residential containers are in the nature of adaptations to collection vehicles that also depart from conventional design.

Commercial wastes are best stored in large metal containers equipped with a hinged lid. The containers range in size from about 2 to more than 30 cubic yards (cu yd). The large containers are designed such that they and their contents can be transported to the disposal site. It is important that arrangement for

**Figure 2.1.** (Courtesy of Zarn, Inc.)

adequate and easily accessible storage facilities be made in the design of commercial buildings. Similar containers are used in multifamily dwellings. An example of a container used in commercial facilities is given in Figure 2.2.

A relatively recent development is the use of containers as ''miniature transfer stations'' in rural areas. The containers are strategically placed in those areas. Problems encountered with the miniature stations usually are traceable to insufficient cooperation on the part of the people benefitting from the service. A diagram showing a miniature transfer station is presented in Figure 2.3. Another relatively simple transfer station using a compactor is shown in Figure 2.4.

Because the practice has ramifications both in storage and in collection, we close this section with a description and discussion of source separation, rather than doing so in the section on collection.

### *Source Separation (Separate Storage and Collection)*

''Separate storage and collection'' in the original sense of the term simply meant separation of garbage (food preparation and kitchen wastes) from rubbish (the other fraction of residential waste). In combined or, as it now is called, ''mixed'' storage and collection, the householder and other waste generators store their waste in the container without doing any separation, hence, the modern designation ''mixed storage and collection.'' The popularity of separation in the original sense of the term soon vanished when circumstances brought an almost abrupt end to the hitherto widespread practice of feeding garbage to swine. Due to human, technological, and economic reasons, mixed collection remained

**Figure 2.2.** Example of container used in commercial facilities. (Courtesy of J.V. Manufacturing, Inc.)

**Figure 2.3.** Small, rural transfer station.

practically the only type of collection practice in the U.S. following the demise of garbage feeding to swine. Indeed, it took the much publicized but nevertheless genuine difficulties besetting conventional solid waste management to transfer the prevailing popularity of combined (mixed) storage and collection to separate storage and collection.

**Figure 2.4.** Transfer station using a compactor.

In recent years, separate storage and collection has reappeared under the name of "source separation." Source separation applies to the segregation of designated recyclables by the householder (or other waste generator) from other recyclables as well as from nonrecyclables. Examples are separation into organic matter suited to biological treatment (composting, biogasification), metallic and plastic containers, glass, "clean" (uncontaminated) paper, newspaper, and the miscellany of waste that remains.

Source separation has acquired a broad appeal for reasons almost too numerous to mention. Indeed, the storage and collection of recyclables is given special attention in the final section of this chapter.

*Arguments Regarding Source Separation.* Although many arguments can be elicited for and against source separation as compared to combined storage and collection regarding their respective economic, technological superiority, and sociological and political appeal, it cannot be contradicted that in practice, recyclables collected through source separation are cleaner (freedom from contaminants) and of higher quality than those reclaimed at a central facility from refuse stored and collected via the mixed route. Theoretically, a mechanical or a combination manual-mechanical system could be designed that would accomplish separation approaching that through source separation. However, the monetary costs would be exceedingly high and the practicality would be dubious.

The technological nature of source separation limits the number and identity of the categories of waste resources that can be separated. The limiting technological aspect is triggered by the need to collect and transport each category of items to its eventual destination and yet maintain the separation throughout. Of course, in a real situation, the actual number of categories is decided by the number of those items for which there is a market and the costs involved in storing and collecting them. However, even if a market were available for each and every category of recoverable resources, the number of categories that could be feasibly accommodated would continue to be rather small.

In terms of storage of source-separated recyclables, container cost is an added expense in that the number of receptacles needed is increased. The cost of a single container large enough to hold a generator's entire daily or weekly output of waste is less than that of the two or more receptacles needed to accommodate that same output when it is split into two or more fractions.

Compartmentalized vehicles designed to maintain the separation are available and are being used successfully. Of course, another approach is to collect the segregated recyclables in separate collection vehicles designed specifically for the purpose. Because of the nature of the recyclables and differences in rate of generation, the frequency of collection of the recyclables could be much less than that of unsegregated refuse which is contaminated with ''garbage.'' Further details regarding the collection of recyclables is given under the heading ''Collection of Recyclables.''

While extolling the advantages of source separation, it should be emphasized that they are contingent upon certain factors, the absence of any one of which could compromise the quality of the reclaimed material. Examples of these factors are sufficient number of separations, proper makeup (nature, composition) of the separations, and last, but not least, cooperation on the part of the waste generator (e.g., householder). An insufficient number of separations results in a disproportionately large mixed fraction and places a limitation on the diversity of the recycling effort. For example, only two separations, one of which is into metal and glass containers, may result in the collection of ''clean'' ferrous or aluminum recyclables, whereas the second separation would be a mixture that only slightly differs from the material collected in ''mixed'' collection. Composition of the individual separations must be compatible with the nature and type of recyclables desired. For example, one of the separations required for obtaining a feedstock suited to biological reclamation should be into organics, i.e., one container should be reserved solely for biologically decomposable organics. If aluminum or ferrous separation is desired, both aluminum objects and ferrous objects can be stored in a common container, since they are readily separated from each other through magnetic separation. However, bimetal containers could pose a problem.

## COLLECTION

This section is concerned almost solely with the collection of residential and commercial refuse, because these materials are the sources of a large fraction, if not most, of the feedstock used in composting and recycling MSW. Another reason is that the collection of domestic and commercial refuse is a rather visible operation. Among the key problems in domestic refuse collection are (1) it is labor intensive, (2) it is vulnerable to traffic conditions, and (3) it is costly.

Innovations aimed at lessening the degree of labor intensity in industrialized countries revolve around reducing crew size from the usual two or three workers to the one-man size, designing the truck to accommodate a one-man crew, and reducing the physical demand on the laborers. With a one-man crew, weight limits on containers and their contents must be strictly enforced. To encourage the cooperation required, the wages of the single crew member must be raised accordingly.

Several municipalities in the U.S. are in the process of evaluating the possibility of switching to either semi- or fully automated collection, or a combination of both. Semi-automated collection requires that the crew member transport (typically roll) the container to the collection vehicle. The vehicle is equipped with a hydraulic device which is designed to engage the container, lift it, and discharge its contents. Conventional packer trucks can be retrofitted with semiautomatic lifting devices. These units cost between $2500 and $3500 (1992).

Fully automated collection, as the term implies, does not require that the crew member come in direct contact with the container. The collector can pick up the refuse container without leaving the vehicle by simply manipulating a boom. The boom is hydraulically activated and generally includes a ''claw'' to grab the container. Obviously, this type of collection requires that the container be placed in an area accessible to the boom (usually at the curb). The diagrams in Figure 2.5 show both semi- and fully automated collection units. Unless the community is extremely homogeneous in terms of topography, street layout, etc., more than likely a combination of semi- and fully automated collection will be required.

In addition, automated collection must be carefully evaluated in terms of cost, compatibility with recycling programs, and type of containers to be used.

Currently, there are about 8 or 10 manufacturers of containers suitable for automated collection in the U.S. A summary of some of the specifications for the containers is presented in Table 2.1. The selection of a particular type of container is an important aspect and should not be taken lightly, since each container costs about $50. In addition, container selection must take into consideration life expectancy (warranty), type of resin, molding process, recyclability, as well as costs associated with assembly and distribution. In early 1992 the costs for assembly were about $1.50/unit and those for distribution were about $1.00/unit.

At the present time, automated collection may be considered in its infancy in the U.S. (Automated collection is much more developed in Europe.) Due to this

**Figure 2.5.** Automated collection units. (Top) Fully automated back-up system (courtesy of Perkins Manufacturing Co.); (bottom) semiautomated collection device (courtesy of Zoeller Waste Systems).

fact, there are a number of important aspects that have not been standardized, such as tests and specifications. Performance tests (drop, capacity, tilt force, stability due to wind force) are just being developed. Selection and acquisition of containers is a rather complex process and is beyond the scope of this book.

Problems with traffic are inevitable because collection involves the use of surface vehicles traveling on public thoroughfares. Adjustment of travel sched-

**Table 2.1 Characteristics of Containers for Automated Collection**

| | |
|---|---|
| Size | 32 to 100 gal |
| Weight | 17 to 58 lb |
| Resin type | High-density polyethylene<br>Medium-density polyethylene<br>Crosslink polyethylene |
| Molding process | Injection, rotational, blow |
| Recycled content | 10 to 100% |
| Cost (1992) | $30 to $65 |

ules to minimize competition with other vehicles is only a partial solution.[6,7] Travel time has a significance because time spent on the road extends crew time and ties up an expensive piece of equipment, namely, the collection truck. Other undesirable effects include aggravation of air pollution and increased fossil fuel consumption. Yet another difficulty with collection takes the form of the objection of citizens to the convergence of collection vehicles at the processing site.

## Considerations

Important considerations regarding collection are (1) combined vs separate collection; (2) frequency of collection; (3) point of collection; (4) crew size; (5) pickup density; (6) programming; and (7) equipment. Since the effect of separate collection was discussed in the section on source separation, discussion of the elements to be considered begins with frequency.

### *Frequency*

Among the several factors that determine frequency are these six: type of collection (combined vs separate); volume of generation; composition (nature) of the waste, effect on rate of waste generation, cost, and fly production. Type of collection influences frequency in that if the garbage fraction is stored separately from the rubbish, the garbage should be collected more frequently than the rubbish, particularly in hot and humid climates. The reason is that it is very putrescible and serves as a source of nutrition and a haven for flies in all their life cycle forms. Once-per-week collection of both garbage and rubbish may be viable in areas with cool climates. In this particular case, the garbage fraction must be properly stored. The rubbish, not being contaminated by garbage and being quite stable, can be stored for a relatively long time without causing a nuisance or public health problems. Consequently, the frequency of rubbish collection is a function mainly of rate of generation and costs. Volume of generation is a self-evident factor in that the indicated frequency of collection is one at which the storage facilities are not overloaded. The importance of composition was mentioned in the comments on effect of type of collection. The more readily decomposable (biologically unstable) the waste, the more frequently it should be collected. With the exception of fly production, putrescibility overrides all the other factors in the determination of frequency.

Frequency probably has an effect on rate of waste generation, even though theoretically it would seem that a "steady state" rate of discard should be reached that would be dependent upon income, standard of living, and other factors. A failure of steady state to materialize may be more apparent than real because density rather than volume is increasing. Thus, the experience has been that once-a-week collection could increase the number of man hours required to collect a ton of refuse by as much as one and one half that required for twice-a-week collection.

Frequency of collection has an impact on source-separation and recycling. Furthermore, collection of nonrecyclables and recyclables on the same day has been found to increase public participation in recycling. Frequency of collection must also consider convenience to the user. The more convenient the design is, the higher the participation and cooperation by the householder.

The effect on fly generation perhaps surpasses the preceding factors in importance because of the bearing on public health. The relation between fly generation and frequency of collection rests upon the interruption of the fly life cycle caused by an increase in frequency of collection. Normally, the life cycle of a housefly (*Musca domestica*) takes about 5 days to complete. Moreover, the increased frequency reduces the opportunity for the flies to ovideposit.

### *Point of Collection*

Point of collection refers to the position of the storage receptacle at the time of collection. The two broad classes of point of collection are "backyard" and "curbside." The terms are self-explanatory. It has been estimated that the time required per ton of waste collected in backyard collection can be about 1.7 times that required with curbside collection. In backyard collection, as much as 30% of the collector's time is spent in walking on private property. Not surprisingly the actual times are highly variable.

Most of the relative advantages and disadvantages of curbside collection vs backyard collection are too self-evident to warrant further discussion.

### *Pickup Density*

Pickup density refers to the number of services per mile. Intuitively — and correctly so — one would surmise that the labor requirements per unit weight (man-hr/ton) of refuse would decrease as the pickup density increased up to a certain point. The reason is that the fewer the pickups, the longer the time per pickup. The general experience is that the breakpoint in terms of pickup time occurs at a density of about 15 services per mile. A variable that influences the actual times and the relationship between number and time is topography of the terrain.

### Programming

Programming is a complex operation even for a small community. It involves planning routes and usage of manpower and equipment. The two disciplines, system analysis and operations research, are of special utility in programming. Operations research involves an evaluation of various ways of using man and machines to find the most efficient arrangement. The requirements for sound programming are a collection of accurate data, a firm comprehension of local conditions, and a thorough appreciation of the human elements, e.g., the workforce. Fixed factors to be considered in programming, i.e., not subject to change, are (1) type of refuse, (2) method of disposal, (3) extent of resource recovery in progress or planned, (4) physical layout, and (5) climatic conditions, especially the meteorological conditions.

### Equipment

Generally, environmental and public health considerations demand that the collection vehicles be covered and watertight. For the most efficient utilization, the capacity of the vehicle should be the maximum allowed by street width, type of roadway, and traffic conditions and regulations. Type of waste dictates the type of vehicle. For mixed refuse in general, the conventional packer truck is appropriate. The truck can be modified slightly by adding on a rack for holding bundles of newspapers and a bin to hold aluminum containers. Open bed trucks may be used to haul demolition debris, tree trimmings, and other aesthetically innocuous materials.

Modifications of truck design have been tried that were aimed at minimizing the dependence upon manpower. Examples are automation of picking up and dumping of containers, equipping the truck with a hydraulic boom mechanism that can grasp a container placed on a curb, elevate it, and empty the container's contents into the body of the truck. Equipment for collecting recyclables is described in the section on collection of recyclables.

## TRANSPORT (HAUL)

A prime factor in the economics of collecting wastes is the cost of moving them from the point of generation to that of disposal or of processing. Factors pertaining to this prime factor are equipment, fuel consumption, and labor costs. The magnitude of the individual costs ultimately is one of distance, or perhaps of time. The consequence of the distance factor is that a definite distance exists beyond which the transport of refuse in the collection vehicle becomes economically unfeasible. The time element enters in because the crew, excepting the

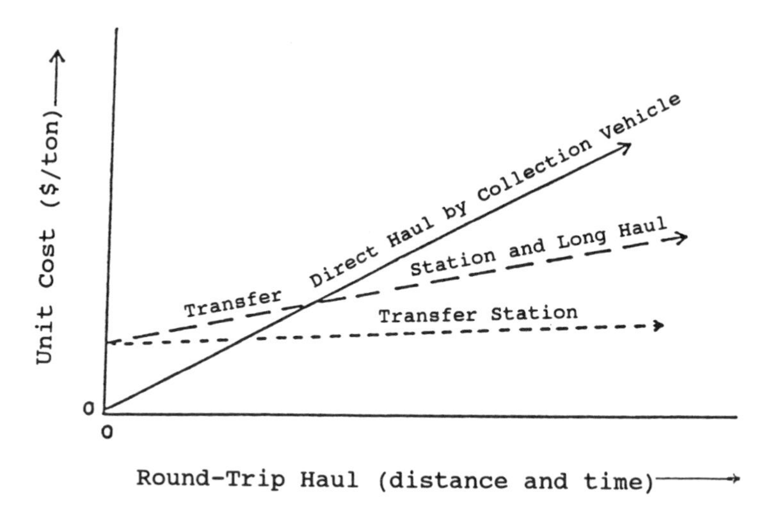


**Figure 2.6.** Relation of direct haul costs to those of a transfer station-long haul combination.

driver, remain idle during the time of transport. Moreover, wear and tear on the collection vehicle is more a function of the length of time the vehicle is in operation than of distance covered. An exception, of course, is tire wear, which is mostly a function of distance traveled, of the speed at which it is traversed, and of the quality of the roads. Time certainly is more inclusive than distance alone, because of the fact that the vehicle is not necessarily continuously in motion nor is the movement at a constant velocity. Consequently, the combined effects of time and distance should be taken into consideration in the determination of an economically practical length of haul in the collection vehicle.

Among the courses to be followed when haul distance extends beyond that economically feasible with the collection vehicle is to transfer the waste in the collection truck to a vehicle (''transfer truck'') that has a capacity significantly larger than that of the collection truck and is manned by a single individual.

Transfer of waste from collection truck to transfer truck may be direct, and in which case, involve only the use of a ramp; or it may be less direct in that the waste is dumped into a receiving hopper or storage pit from which it is loaded on the transfer truck. The transfer facility is known as a ''transfer station''.

In modern waste management, the transfer station should be the site of initial processing, usually in the form of separating potentially useful components from the ''nonuseful.'' Examples are recovery of paper fibers, yard wastes, wood, metals, dirt, and demolition debris. This is done with the objective of reducing the amount of waste destined to be landfilled.

The interrelationship between haul distance and time and cost of haul by collection vehicle and by transfer vehicle is clearly illustrated by the three curves in Figure 2.6. The curve for transfer stations plus long haul does not begin at zero because regardless of distance of haul, the cost of the transfer operation remains constant. As the figure indicates, a "crossover" point exists at which direct haul costs begin to exceed those of the combination of transfer station and long haul. The point at which the two cross is a function of several factors, the more important of which are labor costs (crew size and hourly wage), fuel costs, and traffic.

## STORAGE AND COLLECTION OF RECYCLABLES

The principles and constraints described for storage and collection of mixed MSW are also applicable to source-separated recyclables. These principles and constraints are basic in that they relate to the protection of public health and the quality of the environment. The greater the threat to public health and quality of the environment, the more rigid and extensive become the constraints on storage and collection. Thus, it follows that the constraints applicable to the separate storage and collection of certain types of inherently inoffensive recyclables (e.g., metal and glass containers, office wastepaper, yard waste) are neither as rigid nor as extensive as those required with mixed MSW and food wastes — *provided*, of course, that the recyclables are *uncontaminated* with putrescible waste. On the other hand, constraints on the separate storage and collection of kitchen and restaurant food wastes (i.e., waste generated in the preparation and consumption of meals) and discarded fruits and vegetables in all stages of decomposition are both rigid and extensive.

Systems can be designed for maintaining a desired degree of separation in the storage and collection of recyclables and yet meet the conditions described in the preceding paragraph. The systems specify container type and lists of recyclable materials. At present, the degrees of separation are classified into three broad groups, namely, *mixed waste disposal*, *wet/dry separation*, and *dedicated source separation*.[8]

### Storage

As the name indicates, in *mixed waste disposal* systems, the waste generator stores all solid wastes in the same container. Such systems call for the separation of recyclables to take place at a processing facility. Separation is done either mechanically or manually, or by a combination of the two.

Moisture content serves as the basis for separation in *wet/dry separation* systems. The rationale for such a separation is based upon the applicability of conventional recycling systems to the "dry" fraction and of composting to the "wet" fraction. Thus, moisture content per se is not the ultimate separation characteristic. The terms "wet" vs "dry" are used simply because biodegradable materials generally have a higher moisture content than nonbiodegradable, e.g., discarded plants vs discarded metal or glass containers.

A variation of the wet/dry approach involves the addition of a third separation. In the third separation, recyclables having a high value are separated from those that have a low value. Examples of high-value recyclables are aluminum, cardboard, glass, and plastic containers.

*Dedicated source-separation* systems call for the waste generator to separate his or her solid wastes into particular categories on the basis of recyclability. Wastes not in these categories are subjected to conventional solid waste management. Source-separated recyclables (e.g., glass, aluminum, ferrous) may either be sorted each into a separate container or they may be placed in a single container. The latter form is termed "commingled" source separation. Depending upon the physical and biological characteristics of the recyclables to be stored in them and upon the collection system, types of containers vary from the 6- to 10-gal container used for food wastes (kitchen and restaurant, cf. Chapter 5), stackable plastic containers (generally used in sets of 3), to the conventional 20- to 30-gal container or plastic or paper bag, and eventually to the 90- to 100-gal rolling tote or maybe plastic bags. The final choice depends upon the volume of materials to be collected and the collection system to be used. A sample of containers used in source separation programs are given in Figure 2.7.

## Collection

Among the factors that enter into the selection of a design of a system for collecting recyclables are the following three: (1) the planned degree of source separation; (2) the type of container or containers in which the recyclables are stored at the point of generation until collection; and (3) accessibility to the containers by the collector. Accordingly, collection systems currently in vogue fall into one or more broad categories, namely, *curbside collection*, *containerized collection*, *drop-off collection*, and *buy-back collection*. Typically, all four types are encountered to one degree or another in community-wide recycling undertakings.

With *curbside collection*, separated recyclables are placed on the curb or side of a street for pickup and transfer to a collection vehicle.

*Containerized collection* is the approach commonly used for storing and collecting separated recyclables generated by commercial establishments, public and private institutions, and multifamily residential units. Each building in these establishments is provided with one or more "dumpster"-type containers in which separated recyclables produced in the building are stored for automated collection. Two of the more common types of automated dumpster loaders are the front loader and roll-off loader. Front loaders use steel dumpsters that may have a capacity as large as 8 cu yd. The dumpster is mechanically lifted and dumped over the front of the vehicle. Roll-off dumpsters are large roll-off boxes that may have a capacity as large as 40 cu yd. The filled roll-off dumpster is lifted and placed on the transport vehicle after an empty roll-off has been deposited. Therefore, a roll-off vehicle services only one site per haul. An example of a large container is given in Figure 2.8.

As is indicated by its designation, the distinguishing feature of *drop-off collection* is the placing upon the generator the burden of transporting his or her recyclables to the recycling facility. The facility may or may not have provision for renumerating the generator, i.e., a "buy-back" facility. Storage units used in some drop-off programs are given in Figure 2.9.

### *Types of Collection Vehicles*

One approach to collection is the "co-collection" approach. In one form of co-collection, a conventional MSW packer truck may be used to alternate separate collections restricted to recyclables with separate collections devoted solely to MSW. Alternatively, the definition of co-collection also extends to the concurrent collection in the same truck compartment of MSW and recyclables where at least one of them is contained in bags. Thus, upon discharge of the composite load, MSW and recyclables are available for segregation. Residential MSW collection vehicles are predominantly of two general designs — "rear loading" and "side loading." The standard packer truck has only one compartment. However, this basic design may be modified to provide two or three compartments. An example is presented in Figure 2.10.

Co-collection might be suitable for (1) those situations where a large vehicle fleet exists and cost efficiency dictates the use of fleet for collection of recyclable and mixed waste, and (2) for those operations involving relatively small volume of recyclables, as might be the case with small communities or in the formative stages of a larger operation. A better and the only approach suitable for recycling on a meaningful scale is one that has become known as the "*dedicated collection*" approach. Characteristics of the dedicated approach are (1) adherence to a collection schedule and route not necessarily synchronized with the MSW collection schedule and route; and (2) the collection vehicles are designed and dedicated solely to the collection of recyclables. Vehicles are designed solely for collecting recyclables and are available in a variety of designs. An example is illustrated by Figure 2.11. Recently, vehicles have been designed to serve more than one purpose, as shown in Figure 2.12.

## CAPACITY CONSIDERATIONS

Due to the relatively low densities of most components of solid waste (typically 5 to 20 $lb/ft^3$), collection of materials in the case of recyclables, or of mixed waste in the case of MSW, is necessarily a volume-based undertaking. Thus, in order to optimize collection efficiency and costs, routes and collection vehicle compartment(s) must be properly designed and sized for the composition and quantities of material under consideration. Design of routes and selection of compartment sizes necessitates estimations of quantities of the subject materials (recyclables or MSW). Estimates of quantities are best based upon a thorough waste characterization study. For the majority of current residential recyclables collection programs (i.e., those that collect newspapers and metal and glass containers), set-out quantities can range from approximately 10 to 35 lb per

A

B

**Figure 2.7.** Sample of containers used for storing source-separated materials. (Courtesy of Windsor Barrel Works.)

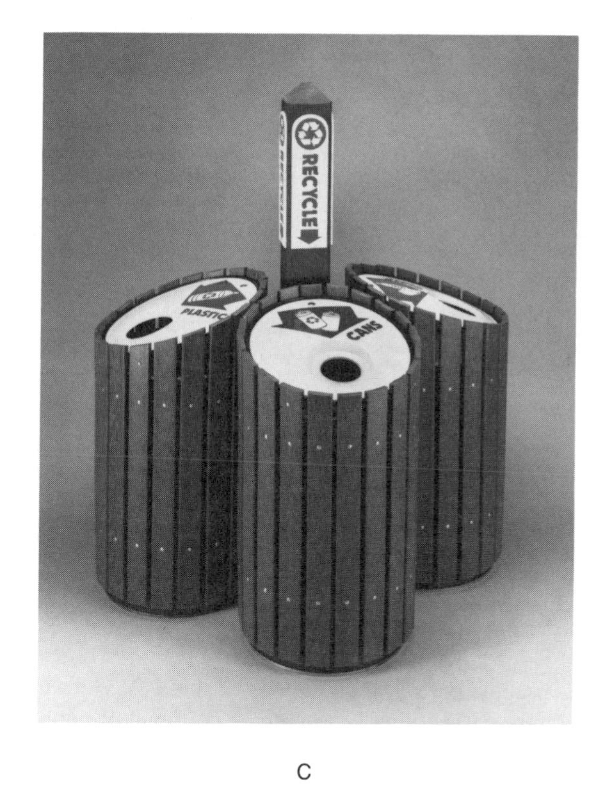


C

D

**Figure 2.8.** Containerized storage at commercial facilities. (Courtesy of J.V. Manufacturing, Inc.)

participating household per week, as shown in Table 2.2. The wide range reflects the variety in the specifics of each program (e.g., level of public education and of promotion, etc.) and the fact that programs with less than weekly collection frequency have been normalized to weekly frequency for the purpose of incorporation of the data in the table. The frequency of collection (e.g., weekly, biweekly, monthly, etc.) influences the average quantities set out over a long-term period. For example, monthly collection of recyclables may result in less capture of recyclables than more frequent collection because of forgetfulness, inconvenience, or nonsynchronization with regular MSW collection, which generally is provided on a once per week basis.

The addition of plastics, i.e., containers manufactured from polyethylene terephthalate (PET), high-density polyethylene (HDPE), or both to curbside collection programs, either at the outset of the program or included later, will add to the total quantities set out. The benefits of including PET and/or HDPE in the program are their relatively high commodity prices and the additional diversion of materials from landfill or incineration. The disadvantages are the relative low as-discarded bulk density of these components (on the order of 1 to 2 lb/ft$^3$); their relatively low percentage in the waste stream (usually 0.2 to 3%); and the lack of commercially proven in-vehicle densification equipment for plastic containers as of this writing. The impact of including plastics in a residential curbside collection program also is illustrated in Table 2.2.

If markets are available, mixed paper may be added to the curbside program. Mixed paper typically constitutes approximately 10 to 20% of residential wastes, thus representing relatively substantial quantities for collection with the attendant volumetric requirements.

**Figure 2.9.** Storage units used in drop-off programs. (Courtesy of Kotrac.)

One of the judgmental factors regarding types and quantities of materials to be collected, and therefore vehicle capacity, is almost always cost, or more specifically for this discussion, anticipated revenues from the sale of recyclables. To form a judgement, the relative contribution of each of the potentially targeted recyclables is an important aspect. Two methods of representing the commodity revenue contributions are illustrated in Tables 2.2 and 2.3.

The specific recyclable components selected for collection and their estimated rates of generation and set out are the key variables in the determination of collection frequency and compartment capacity. Lack of attention to detail and

**Figure 2.9 (continued).**

to wellfounded estimations results in the premature filling of one recyclable compartment to capacity while other compartments are underutilized. The newspaper compartments and plastic compartments are two that historically have been underestimated in terms of adequate volume. On a mass basis, newspapers and plastics typically represent a range of 50 to 70% and 2 to 4%, respectively, of recyclables placed out for collection, as shown in Table 2.3.

Trailers are available with fixed or adjustable compartments, and either are equipped with a self-dumping mechanism or the compartments are removable using a fork-lift or similar equipment. In the latter case, double handling decreases the overall efficiency of the collection system. End-dump and side-dump designations for collection vehicles refer to the direction of automated unloading with respect to the longitudinal axis, or the cab end, of the chassis. Thus, end-dump refers to vehicles that discharge their loads from the rear of the vehicle. Side-dump vehicles discharge to the left, right, or both. Combination-dump vehicles have certain compartments that dump to either side and a rear compartment that discharges at the rear of the vehicle. Compartment volumes

**Figure 2.10.** Collection vehicles for recyclable materials. (Courtesy of Leach Company.)

**Figure 2.11.** Side loader vehicle with tipping mechanisms. (Courtesy of Kann.)

**Figure 2.12.** Dual-purpose collection vehicle. (Courtesy of Oshkosh Truck Corporation.)

typically range from 25 to 35 cu yd for trailer, end-dump, and side-dump vehicles. An example of an end-dump vehicle is given in Figure 2.13.

Regardless of method of categorization, fundamental issues that impact effective utilization of collection labor include lifting height, reach, and steering wheel location (i.e., right, left, both sides of the cab). The potential implications of some of these issues are illustrated in Table 2.4 as an example. The example

**Table 2.2 Information Typical of Residential Curbside Collection Programs**

| | Commingled Material Category | |
|---|---|---|
| | Metal and Glass Containers, Newspapers | Plastics (HDPE and PET) |
| lb/hh/wk[a] | 4 to 25 | 0.08 to 0.20 |
| Participation (%) | 40 to 70 | 20 to 25 |
| lb/pph/wk[b] | 10 to 35 | 0.3 to 0.9 |
| Market price ($/ton material category) | 30 to 50 | 100 to 200 |

[a] Pounds per participating and non-participating households (hh) per week.
[b] Pounds per participating household (phh) per week.

**Table 2.3 Typical Ranges of Composition and Commodity Revenues for Residential Curbside Programs**

| Material | Weight (%) | Revenue Contribution (%) |
|---|---|---|
| Newspaper | 50–70 | 35–45 |
| Glass containers | 20–30 | 10–20 |
| Aluminum cans | 1–2 | 35–45 |
| Steel cans | 5–10 | 1–4 |
| PET and HDPE containers[a] | 2–4 | 5–15 |

[a] PET = polyethylene terephthalate;
HDPE = high density polyethylene.

**Figure 2.13.** "End-dump" vehicle. (Courtesy of Kann.)

considers the case of one driver operating the collection vehicle. For a specific activity, the differences in time among the columns represent cost savings or expenditures among the collection vehicle alternatives that are listed in the table. When taken collectively over a round trip, the time and cost savings attributable to efficient use of labor and motion can be substantial. An example of the results

**Table 2.4 Residential Curbside Collection Time Loss Example**

| Activity and Time Loss Variable | Trailer System | Lo-Profile[a] End-Dump | Lo-Profile Side-Dump |
|---|---|---|---|
| Dumping mechanism | Forklift | Hydraulic lift | Hydraulic lift |
| Time lost/trip | 15 to 25 min | 8 to 12 min | 8 to 12 min |
| Loading recyclables | Left side | Right side | Right side |
| Time lost/stop | 10 or 15 sec | 0 | 0 |
| Loading height | | | |
| Time lost/stop | 3 to 6 sec | 0 | 0 |
| Braking Mechanism | | | |
| Time Lost/Stop | 2 to 3 sec | 2 to 3 sec | 0 |

[a] Lo-Profile refers to those vehicles designs where the chassis is situated low to the ground to reduce the height of lifting by the worker.

**Table 2.5 Residential Curbside Collection Vehicle Evaluation (70 TPD)**

| | Trailer System | Lo-Profile End-Dump | Lo-Profile Side-Dump |
|---|---|---|---|
| Vehicle costs (each) | $30,000 | $60,000 | $85,000 |
| Annual costs | | | |
| Vehicle debt service ($) | 86,000 | 147,000 | 156,000 |
| Labor ($) | 637,000 | 418,000 | 380,000 |
| Maintenance and supplies ($) | 313,000 | 235,000 | 214,000 |
| Total system costs ($) | 1,036,000 | 633,000 | 591,000 |
| Unit collection and unloading cost ($/ton) | 57 | 44 | 41 |

of the time and motion analysis presented in Table 2.4 is shown in Table 2.5 for the same collection vehicle alternatives. In this example, the unit collection and unloading cost is 40% different between the highest and least cost alternatives. The cost of collection can range from about $10 to more than $100 per ton. The wide range of costs is due to a number of factors including type of vehicle, method of collection, type and number of containers, and design of the system.

## MATERIAL HANDLING ISSUES

Collection is fundamentally an exercise in material handling; as such, a key issue is the optimization of the time and motion involved in the activity. As mentioned previously, many factors influence collection frequency and the time that must be dedicated to the collection and transporting of materials. One important factor is the types of collection vehicles for the task at hand. Currently, the types of collection vehicles are virtually endless, and it is likely that over the next several years the top performers will become apparent. However, at this point in time, only a general categorization is prudent. One form of categorization is presented in Table 2.6.

**Table 2.6 General Types of Recyclables Collection Vehicles**

| Main Category | Advantages | Disadvantages |
|---|---|---|
| Trailer | Use with existing MSW collection vehicles, low capital cost | Large turning radius, requires separate motive equipment |
| End-dump | Has own motive source, unidirectional unloading, medium capital cost | No concurrent unloading of compartments possible (compartments are emptied consecutively) |
| Side-dump | Concurrent unloading of compartments possible | High capital cost |
| Combination-dump | Concurrent unloading of compartments possible | High capital cost |

## REFERENCES

1. Diaz, L.F., G.M. Savage, and C.G. Golueke, *Resource Recovery from Municipal Solid Wastes* , Vols. I and II (Boca Raton, FL: CRC Press, Inc., 1982).
2. Tchobanoglous, G., Thiessen, H., and Eliassen, R., *Solid Waste Engineering Principles and Management Issues* (New York: McGraw-Hill, 1977).
3. Zandi, I. and J.A. Hayden, "The Flow Properties of Solid Waste Slurries," *Compost Sci. J. Waste Recycl.*, 12(3):18–30, March-April (1971).
4. Browning, J.E., "Garbage — Pipeline Program," *Chem. Eng.*, 78(17):60–62, July 26 (1971).
5. Reindl, J., *Rural Solid Waste Collection Systems,* G2853, Cooperative Extension Programs, University of Wisconsin Extension, Madison, March (1977).
6. Bodner, R.M., E.A. Cassell, and J.P. Andros, "Optimal Routing of Refuse Collection Vehicles," *J. Sanit. Eng. Div. ASCE,* 96(SA4):893 (1970).
7. Liebman, J.C., "Systems Approaches to Solid Waste Collection Problems," paper presented at the Meet. A.A.A.S., Chicago, December (1970).
8. *Recyclables Collection: Universe of Alternatives,* Preliminary Draft, prepared by CalRecovery, Inc. for the New York City Department of Sanitation, April (1991).

# CHAPTER 3

# Waste Characterization

## INTRODUCTION

An essential preliminary step in municipal solid waste (MSW) management is the accurate determination of the quantity and characteristics of the feedstock of concern, namely MSW. The applicable characteristics include physical, chemical, and thermal properties as well as quantity and composition. Reasons for need and procedures for making the determination are presented in this chapter.

Although the advisability and necessity for acquiring the quantitative and qualitative knowledge regarding the waste stream to be managed are indisputable, until recently the literature has been singularly devoid of descriptions and discussions of practical procedures for obtaining that knowledge. Having referred to the need for determining the characteristics, we proceed to describe the four major classes of solid waste and the components of wastes in terms of their general recyclability and their general amenability to biological treatment. Thereafter, we propose ways and means of acquiring that information, concentrating on information that pertains to composting and to recycling. We close the chapter with a description and discussion of candidate methods for determining quantities, composition, and other properties of the waste.

### Need for Accurate Determination of Waste Characteristics

The development of a rational comprehensive strategy for waste management rests upon the acquisition of sound information on the quantity, generation rate, and the pertinent properties of the waste to be managed. This requirement not only is applicable to waste management taken in its entirety, but also to the individual constitutive elements of the overall plan. For example, if composting (or biogasification) is one of the elements in the overall plan, knowledge of the quantity and characteristics of the biodegradable fraction of the waste stream provides information on the feasibility of the projected operation; on the design and required capacities of processing equipment and on pertinent operational features; and on the likely characteristics of the recovered product(s). The same reasoning applies to the development of a recycling program. In this case, recyclable materials are the fraction of interest.

Obviously, the determination of both current and future waste characteristics are required for planning solid waste management systems. Consequently projections and estimations of future quantities and characteristics are important. The projections and estimations require many judgments, including those of buying habits, product recyclability, and growth of business and population.

### Relation of Disposed Waste, Diverted Waste, and Generated Waste

Within the U.S. and as recently as the 1980s, the attention of solid waste management was predominantly on landfill disposal of wastes. For the purpose of discussion, wastes destined to landfill disposal are defined herein as "disposed" wastes. With the advent and expansion of alternatives to landfill disposal (e.g., recycling and composting programs), the quantities of generated wastes that formerly would have been landfilled grew to substantial levels and the accounting of these quantities became necessary and important. The wastes diverted by these diversion programs are defined herein as "diverted" wastes. The summation of the quantity of disposed waste and of diverted waste is the total quantity of generated waste.

Wastes that are reduced in volume and mass due to thermal, chemical, and biological reactions (e.g., direct combustion, pyrolysis, anaerobic digestion, etc.) in some cases are accounted for as diverted wastes (i.e., diverted from landfill disposal except for the ash residue) and in other cases as disposed wastes. No standardization on this matter currently exists. The important aspect is that the definitions be consistent, uniformily applied, and that the quantities be properly accounted for. From the relation among disposed, diverted, and generated wastes it follows that in areas where the quantity of diverted wastes are zero or very small in relation to that of disposed wastes, the quantity of generated wastes is, respectively, equal to or approximately equal to the quantity of disposed waste.

Historically, accounting for disposed waste was relatively easy. The reason is that disposed wastes were by far the majority of wastes that were generated by a community and that these wastes were taken in large quantities to solid waste facilities. In general, the situation for diverted wastes was and is currently much more complex. The reasons are that generally diversion occurs (or can occur) in many different forms, for a number of material types, among many collection companies, and among many end users. Thus, in very few instances can diverted wastes from a specific community be weighed in large quantities at one location. The upshot is that the accurate measurement of quantities of diverted wastes by material category and by community is usually inconvenient and difficult. The advent of relatively large capacity centralized processing facilities for diverted wastes somewhat ameliorates but by no means eliminates the foregoing situation. The development of cost-effective and accurate methods for determining diverted quantities within communities is needed in the solid waste industry.

## MAJOR CLASSES OF MSW

### Traditional

Traditionally, MSW was classified into three major classes based on source of generation, i.e., into residential, commercial, and industrial. Based on that classification, residential, commercial, and industrial solid wastes consisted of garbage and rubbish. The garbage fraction was primarily in the form of waste associated with the preparation and consumption of food (e.g., meat and vegetable scraps) often called putrescibles. All wastes not classified as ''garbage'' were designated ''rubbish.'' Thus, classically the mixture of garbage and rubbish were known as refuse (MSW). Major components of rubbish were glass, metal, and plastic containers, packaging material, waste paper and paper products, and yard and garden debris. Major components of rubbish in developed countries also included wood waste and discarded appliances. The data in Table 3.1 typify the traditional classification of waste composition as well as other general characteristics.

### Current

The classification of sources of generation currently in vogue is much the same as the classical, except that a separate class of generation, self-haul, has been defined. Self-haul waste is waste delivered by the public and small private haulers to solid waste facilities. Although self-haul wastes are in reality a subset of the major classes (predominantly residential and commercial sources of generation), the fact that self-haulers can commonly constitute 15 to 30% of deliveries to the solid waste facilities is reason to justify the separate classification.

The waste components currently in use are more extensive than those given in Table 3.1. The reason is that solid waste practices require comprehensive, fundamental, and detailed knowledge of materials in the waste stream. Two of the stimulations in the need for comprehensive information are the control of product quality and the control of pollution from solid waste facilities. A listing of components that reflects major component categories for which data are available across the U.S. is given in Table 3.2. Definitions of the major component categories are provided for reference in Table 3.3.

As was true in the past, currently the components that comprise the composition of the MSW stream reflect the life style and prosperity of the population and the care with which the municipality manages the solid waste. For example, extensive residential and commercial construction activity contributes disproportionate quantities of construction debris to the MSW stream. In many developing countries, coal and wood are commonly used in the home as fuel in food preparation and heating. Not surprisingly, the MSW generated in those countries contains large amounts of ash, as is shown by the data in Table 3.4.

**Table 3.1 Quantity and Composition of Municipal Solid Waste[a] (Percent Wet Weight Unless Otherwise Noted)**

| Material Category | India, Urban | Manila, Philippines | Asuncion, Paraguay | Lima, Peru | Mexico City, Mexico[2] |
|---|---|---|---|---|---|
| Putrescibles | 75.0 | 48.8[b] | 60.8[b] | 34.3 | 56.4[b] |
| Paper | 2.0 | 17.0 | 12.2 | 24.3 | 16.7 |
| Metals | 0.1 | 1.5 | 2.3 | 3.4 | 5.7 |
| Glass | 0.2 | 5.3 | 4.6 | 1.7 | 3.7 |
| Plastics, rubber, leather | 1.0 | 6.5 | 4.4 | 2.9 | 5.8 |
| Textiles | 3.0 | 3.7 | 2.5 | 1.7 | 6.0 |
| Ceramics, dust, stones | 19.0 | 17.2 | 13.2 | 31.7 | 5.7 |
| Weight/capita/day (lb) | 0.91 | 0.88 | 1.41 | 2.12 | 1.50 |
| Average bulk density (lb/ft$^3$) | — | 13 | 24 | 11 | — |

| Material Category | Caracas, Venezuela | Sasha Settlement Ibadan, Nigeria | Berkeley, CA U.S. | Broward Co., Florida U.S. |
|---|---|---|---|---|
| Putrescibles | 40.4[b] | 76.0 | 39.0[c] | 39.8[c] |
| Paper | 34.9 | 6.6 | 40.1 | 37.8 |
| Metals | 6.0 | 2.5 | 3.0 | 5.6 |
| Glass | 6.6 | 0.6 | 7.6 | 6.7 |
| Plastics, rubber, leather | 7.8 | 4.0 | 6.3 | 9.0 |
| Textiles | 2.0 | 1.4 | 1.7 | — |
| Ceramics, dust, stones | 2.3 | 8.9 | 2.3 | 1.1 |
| Weight/capita/day (lb) | 2.07 | 0.37 | 3.10 | 3.86 |
| Average bulk density (lb/ft$^3$) | 14 | — | — | — |

*Source:* Diaz and Golueke[1,2] and CalRecovery, Inc.[4]

[a] Based on actual measurements.
[b] Includes small amounts of wood.
[c] Includes yard wastes, food wastes, and wood.

**Table 3.2 List of Some Major Waste Component Categories and Subcategories**

| | |
|---|---|
| Mixed paper | Other organics |
| High-grade paper | Ferrous |
| Computer printout | Cans |
| Other office paper | Other ferrous |
| Newsprint | Aluminum |
| Corrugated | Cans |
| Plastic | Foil |
| PET bottles | Other aluminum |
| HDPE bottles | Glass |
| Film | Clear |
| Other plastic | Brown |
| Yard waste | Green |
| Food waste | Other inorganics |
| Wood | |

**Table 3.3 Description of Some Major Waste Component Categories**

| Category | Description |
|---|---|
| Mixed paper | Office paper, computer paper, magazines, glossy paper, waxed paper, other paper not fitting categories of "newsprint" and "corrugated" |
| Newsprint | Newspaper |
| Corrugated | Corrugated medium, corrugated boxes or cartons, brown (kraft) paper (i.e., corrugated) bags |
| Plastic | All plastics |
| Yard waste | Branches, twigs, leaves, grass, other plant material |
| Food waste | All food waste except bones |
| Wood | Lumber, wood products, pallets, furniture |
| Other organics/ combustibles | Textiles, rubber, leather, other primarily burnable materials not included in the above component categories |
| Ferrous | Iron, steel, tin cans, bimetal cans |
| Aluminum | Aluminum, aluminum cans, aluminum foil |
| Glass | All glass |
| Other Inorganics/ noncombustibles | Rock, sand, dirt, ceramics, plaster, nonferrous nonaluminum metals (copper, brass, etc.), bones |

**Table 3.4 Average Composition of Residential Waste in Seoul, Korea**

| | Concentration (Percent Dry Weight) | |
|---|---|---|
| **Component** | **Summer** | **Autumn** |
| Paper | 16.17 | 3.51 |
| Wood | 7.44 | 0.74 |
| Textiles | 3.84 | 1.33 |
| Food waste | 10.29 | 3.56 |
| Plastics | 5.98 | 2.09 |
| Leather/rubber | 3.63 | 0.13 |
| Other organics | 4.54 | 2.64 |
| Subtotal organics | 51.89 | 14.00 |
| Ferrous metals | 3.67 | 0.26 |
| Non-ferrous metals | 0.46 | 0.08 |
| Glass | 10.60 | 0.68 |
| Ceramics, stones | 10.30 | 1.54 |
| Other inorganics | 2.54 | 2.31 |
| Briquet ash | 20.54 | 81.13 |
| Subtotal inorganics | 48.11 | 86.00 |
| Total | 100.00 | 100.00 |

*Source:* Japan International Cooperation Agency.[6]

The data in the table indicate that the concentration of ash in residential waste generated in Seoul, Korea, during the summer is about 21%. The concentration increases to more than 81% in the autumn. The negative impact of the ash on resource recovery activities in developing countries has now been recognized, and efforts are being made to store and collect the ash separately.[1,2]

A determination of the composition of wastes for particular subclasses of generators provides the characteristics of specific generators as well as enables the identification of opportunities for resource recovery. Subclasses of generators worth characterizing include single family, multifamily, schools, office buildings, and large local businesses and industries. An example of this manner of characterization is illustrated in Table 3.5.

## Components of Solid Waste

This section discusses some of the components of wastes that are produced by the sources that were discussed in the preceding section. Components selected for attention are those that are of particular concern in biological treatment systems and in recycling. Components include tin cans, newspapers, yard waste, wood wastes, glass containers, textiles, etc. Synonyms in the industry for components are material categories, types, and constituents.

### *Significance of Components*

The components of solid waste to be treated are a fundamental determinant of the technical and economic feasibility of composting or of biogasification as a treatment option or of recycling as a method of resource recovery. The nature and composition of a waste determine its biological, chemical, and physical characteristics in terms of potential suitability as a feedstock for biological processing or for a recycling program. Nature and composition also determine the extent and manner of the processing that may be involved in putting the potential into practice. Moreover, the characteristics of the components not only determine suitability, they also determine to a large extent the quality of the compost product or of the recycled components.

Some waste components may have chemical constituents that could be hazardous to personnel and to equipment. A constituent may represent a hazard because it is plant or animal pathogen, or is toxic or highly corrosive or extremely flammable, or dangerously explosive.

The U.S. EPA has promulgated a definition of hazardous wastes that is appropriate for both developed and developing countries. According to the definition, a waste is hazardous if it poses a substantial present or potential hazard to human health or living organisms because the waste is nondegradable or persistent in nature or it can be biologically magnified, or it can be lethal, or the waste may otherwise cause or tend to cause detrimental cumulative effects. Inasmuch as it is likely that the harmful constituent would not be affected by biological treatment, it would become a part of the compost or sludge product. A consequence would be a serious reduction in the safe use of the product and, hence, in its utility.

## Some Special Types of Waste

The successful structuring of waste characterization programs and successful acquisition of adequate and accurate data require a basic knowledge and understanding of waste processing and its terminology if recovery of materials, energy, or both is the objective.

**Table 3.5 Average Waste Composition of 11 Generator Subclasses (Wt %)**

| Component | Front Loaders | Rear Loaders | Multifamily Dwellings | Hotels | Schools | Office Buildings | Shopping Centers | Durable Goods Warehouses | Electronics Mfrs. | Paper and Wood Prods. Mfrs. | Auto Dealers/ Repair Shops | Plastic Products Mfrs. | Construction |
|---|---|---|---|---|---|---|---|---|---|---|---|---|---|
| Mixed paper | 12.1 | 10.5 | 10.3 | 11.3 | 28.6 | 43.0 | 11.8 | 12.9 | 15.9 | 8.5 | 19.8 | 13.1 | 2.2 |
| Newsprint | 13.3 | 18.0 | 33.7 | 11.5 | 2.5 | 3.5 | 6.7 | 7.2 | 13.2 | 19.8 | 1.7 | 0.4 | 0.5 |
| Corrugated | 20.0 | 9.6 | 10.0 | 27.5 | 26.1 | 21.5 | 37.4 | 35.6 | 44.9 | 10.3 | 29.4 | 27.8 | 17.4 |
| Plastic | 8.7 | 9.1 | 8.6 | 13.7 | 6.9 | 3.7 | 8.0 | 11.1 | 5.6 | 2.0 | 7.3 | 22.2 | 3.8 |
| Yard waste | 5.5 | 22.5 | 5.3 | 0.5 | 0 | 0.0 | 2.8 | 0.5 | 0.7 | 3.2 | 3.6 | 0.0 | 3.2 |
| Wood | 10.3 | 2.8 | 1.7 | 0.7 | 0.1 | 8.5 | 3.6 | 15.0 | 9.4 | 46.6 | 0.5 | 27.2 | 27.1 |
| Food waste | 9.1 | 8.0 | 13.0 | 4.9 | 6.7 | 2.2 | 12.4 | 0.7 | 0.2 | 0.2 | 1.6 | 0.4 | 1.0 |
| Other organics | 5.4 | 6.1 | 7.4 | 7.5 | 0.1 | 1.2 | 1.9 | 2.0 | 0.2 | 5.8 | 8.3 | 2.2 | 2.3 |
| Total combustibles | 84.4 | 86.6 | 90.2 | 77.6 | 71.0 | 83.6 | 84.4 | 85.0 | 90.1 | 96.4 | 72.2 | 93.3 | 57.5 |
| Ferrous | 5.3 | 4.4 | 2.5 | 6.9 | 8.2 | 0.6 | 3.6 | 10.0 | 3.9 | 0.6 | 16.4 | 0.8 | 17.5 |
| Aluminum | 1.2 | 1.1 | 1.1 | 2.7 | 1.5 | 0.4 | 0.7 | 0.4 | 0.3 | 0.1 | 1.5 | 0.9 | 0.1 |
| Glass | 5.3 | 6.8 | 5.9 | 11.9 | 6.2 | 0.8 | 2.7 | 1.5 | 1.4 | 0.1 | 9.9 | 4.9 | 1.0 |
| Other inorganics | 3.8 | 1.1 | 0.3 | 0.9 | 13.1 | 14.6 | 8.6 | 3.1 | 4.3 | 2.8 | 0 | 0.1 | 23.9 |
| Total noncombustibles | 15.6 | 13.4 | 9.8 | 22.4 | 29.0 | 16.4 | 15.6 | 15.0 | 9.9 | 3.6 | 27.8 | 6.7 | 42.5 |
| Total | 100 | 100 | 100 | 100 | 100 | 100 | 100 | 100 | 100 | 100 | 100 | 100 | 100 |
| Vehicles sampled | 51 | 38 | 9 | 5 | 2 | 5 | 10 | 8 | 6 | 4 | 3 | 4 | 9 |

*Source:* CalRecovery, Inc.[7]

### *Processible Wastes*

In terms of potential utility, most of the components in the MSW stream are suitable candidates for composting and recycling. Unfortunately, however, without intervention these suitable components are mixed with nonrecyclable components and discarded as mixed MSW. The exception is recyclable components that are separated from other wastes at the site of generation, and collected separately, i.e., source-separated materials. Wastes or components that are amenable to processing and recovery are termed processible wastes. Wastes that cannot be accommodated by the processing system are defined as nonprocessible wastes.

Ideally only those wastes for which a given facility has been specifically designed should be accepted and processed by the facility. An exception might be a component that has been shown to fit within the existing or appropriately modified design of the facility and has the appropriate biochemical characteristics or recycling potential.

The reality is that the ideal situation rarely exists, and accordingly, nonprocessible materials contaminate the otherwise processible wastes. The extent of the contamination varies from relatively low in the case of collecting and processing source-separated materials to relatively high in the case of mixed MSW. In the latter case, nonprocessible wastes can approach 5 to 10% of the total depending on the conditions, thus the importance and relevance of the concept of nonprocessible and processible wastes to waste characterization.

### *Institutional Wastes*

Wastes generated by institutions such as hospitals, medical clinics, schools, and public (governmental) agencies are sometimes classified as "institutional" wastes. With the exception of the medical (infectious) fraction, institutional wastes are not hazardous nor do they constitute a handling problem. All medical waste should be contained in special, closed, plastic bags and should be collected and disposed of separately, preferably by incineration or by special processing and handling in a sanitary landfill.

### *Sewage Sludge*

Although the treatment and disposal of sewage sludge is not within the scope of this book, some space is devoted to the subject not only because sewage sludge is a municipal waste, but also because composting has become a leading method for treating sewage sludge.

Sewage sludge is the concentrated settleable solids fraction of wastewater (sewage) that has been subjected to some form of treatment. Forms of sewage treatment range from those that are quite simple to those that often are very complex. The simplest form of treatment involves only the settling of solids that takes place in residential and community septic tanks and in the primary stage of conventional treatment. Sludge thus produced is termed "raw" sewage sludge. The more complex forms of wastewater treatment involve the subjection of the sludge and of the supernatant from the primary stage to further processing (i.e., secondary and tertiary treatment). A common form of sludge treatment is anaerobic digestion. Because sludges produced in primary treatment as well as in further treatment generally have a settleable solids concentration ranging from a little less than 1 to 4%, they usually are dewatered to a solids concentration of about 15% or higher before being subjected to final treatment or to final disposal.

Due to the attendant hazard posed to the public health and especially to workers' health, raw (primary) sewage sludge should not be accepted by an MSW compost facility. The hazard is in the dense concentrations of human pathogens in the raw sludges. If for some compelling reason it must be composted, it should be done so at a separate facility and with the imposition of all necessary safeguards. Although composting will minimize and perhaps eliminate the hazard from pathogens, it would have no effect on the heavy metals and only partially on resistant toxic organics. Heavy metal or toxic organics concentrations higher than trace level may place a particular sludge in the hazardous waste category and certainly would place serious constraints on the use and disposal of the sludge.

### *Low Bulk Density Waste*

Although source-separated waste of low bulk density such as bulk plastic containers, synthetic fibers, foam, plastic film, and corrugated may have a recycling potential, they pose potentially significant problems during collection, processing, or both. The potential problems include inordinate volume requirements for storage onboard collection vehicles and for storage at the processing facility.

## PROCEDURES

### Importance of Procedure Selection

In the introduction to this chapter, the need for accurate assessment of the quantity and characteristics of MSW was discussed on a general plane. In this section, the emphasis is specifics and involves evaluations and comparisons of current procedures for determination of waste quantities and composition.

#### *Background*

As was pointed out previously in this chapter, the quantity and composition of MSW vary with location. This variation is due to the fact that generation rate and composition are affected by several factors, not the least of which are degree of industrialization, socioeconomic development, and climate.

Despite the variations, four general trends in the variation can be detected in an individual country, at least between similar types of neighborhood in the cities. The existence of trends is not surprising in view of the similarities between the socioeconomic and physical conditions that often exist among the regions of a given country. The first trend — related to quantities — suggests that per capita waste generation is proportional to the degree of prosperity as indicated by level of per capita income. The second trend is an increase in the use and discard of paper wastes commensurate with national advance in development. The third trend is a drop in percentage of putrescible materials in municipal wastes. The third trend generally follows as a consequence of the second trend and as a consequence of advances in technical and economic development. An example of the latter is the use of garbage disposal equipment in residential dwellings, thus diverting food wastes from the solid waste stream to the wastewater treatment system. The fourth trend is an increase in density of refuse with decrease in per capita income (e.g., from 10 to 24 $lb/ft^3$). The increase is a result of a lower concentration of paper and paper products and a larger proportion of kitchen wastes (which are high in moisture content), ashes, and dirt.

Recently, at least in the U.S., the promotion and implementation of waste minimization programs, waste prevention programs, and onsite reuse of waste materials has begun to influence the rate of generation and therefore the rate of discard. Thus, the properties of the waste stream will also be influenced by the above waste management strategies.

## Recommended Study Procedures

To account for all the waste disposed within the survey area, it is essential to ensure that all disposed waste in the area is weighed. A similar goal applies in the case of diverted waste, but difficulties exist as described previously.

### *Quantity*

Not surprisingly, experience shows that an accurate determination of disposed waste quantities depends upon data obtained by way of a properly conducted gravimetric study (i.e., weight survey). Furthermore, experience also shows the inadequacy of volumetric (as opposed to gravimetric) determinations, and hence, that they are an unsatisfactory substitute for an in-depth gravimetric study. The usual method of volumetric determination (or survey) consists of recording the number of loaded vehicles entering the solid waste facility (e.g., transfer station or landfill) during a given time interval and multiplying the number by the product of the assumed volumetric capacity of the individual vehicles and by an average in-vehicle density of waste. In the usual method of volumetric determination the density of waste is not rigorously determined during the course of the survey. Typically a density value is extracted or inferred from the literature. Usually the only accurate number obtained by means of the vehicle count approach is that of the vehicles entering the facility.

At least two factors can be advanced as reasons for the usual inadequacy: (1) not every vehicle is fully loaded; that deficiency precludes volumetric accuracy; and (2) gravimetric accuracy is precluded by the failure to account for density of the wastes during the course of the survey. In some cases the volumetric data are converted to weight data using an assumed "average" density from the literature. Needless to say, this procedure is inappropriate and inaccurate unless the literature value is confirmed via a properly constructed study. An additional factor pertains only to certain developing countries. With respect to equating disposed waste to waste generation in developing countries, a precaution is that in those countries, collection trucks may dump their loads at an unauthorized site or substantial scavenging may take place en route, and no quantitative accounting is made of these pre-disposal activities.

At present, the only procedure for arriving at an accurate estimate of the quantity of disposed waste is to weigh each vehicle and its load as it enters the treatment facility. The vehicle and contents are weighed by means of a weighing scale that can accommodate vehicles of any size that come to the site. Of course, tare weight must be determined. (Tare weight is the weight of the empty vehicle.) The measurements should also include loads delivered by private individuals and small private contractors (self-haul).

### *Scales and Weighing Protocol*

Several types of suitable scales are on the market. The selected scale may be one that is permanently installed, or a portable version may be used. Available at present are models that are both convenient and sufficiently accurate. They are equipped with load cells that can be powered either by direct or by alternating current. Also available are other types that do not require an outside power source. Some large metropolitan areas in developing countries have industries (such as cement, chemical, etc.) that commonly own truck scales which they may be willing to make available for the survey. An example data sheet for a weight survey is reproduced in Figure 3.1.

If it is not feasible to weigh each and every incoming loaded vehicle, then a procedure may be substituted that entails the weighing of a sample of randomly selected loaded vehicles to determine an average weight of waste per vehicle. To arrive at the total input, the average weight of waste per vehicle is multiplied by the total number of loads over a specified time period. Although survey results obtained by this modified method may not be as accurate as those obtained by weighing each vehicle, they are better than data obtained without having resorted to any actual weighing. The accuracy can usually be improved if the data are collected separately for residential, commercial, industrial, and self-haul sources. This method is particularly adaptable and useful for determining quantities of self-haul and waste delivered in roll-off containers.

A third method, which also is not as accurate as the first method, is a variation of the second method and is of similar accuracy. The third method involves the collection of the following data: (1) average density of delivered wastes;

Date: ____________ Recorded by: ____________

| Type of Generator | | | Weight/Volume | | | | Self-haul Wastes | | | | | | | | | | | |
|---|---|---|---|---|---|---|---|---|---|---|---|---|---|---|---|---|---|---|
| | | | | | | | Self-haul Vehicles | | | | | | | | Waste Categories | | | |
| Resid. | Comm. | Indust. | Incoming Gross Wt.(lbs) | Tare Wt.(lbs) | Estimated Volumetric Capacity of Vehicle (yd3 | Estimated Volume of Waste (%) | Mini-pickup | Full-Siz pickup | Dump Truck | Flatbed Truck | Trailer | Car Trunk | Station Wagon | Van | Yard Waste | C & D | Dirt/ Rubble | Misc. |
| | | | | | | | | | | | | | | | | | | |
| | | | | | | | | | | | | | | | | | | |
| | | | | | | | | | | | | | | | | | | |
| | | | | | | | | | | | | | | | | | | |
| | | | | | | | | | | | | | | | | | | |
| | | | | | | | | | | | | | | | | | | |
| | | | | | | | | | | | | | | | | | | |
| | | | | | | | | | | | | | | | | | | |
| | | | | | | | | | | | | | | | | | | |

**Figure 3.1.** Sample data sheet for a weight survey.

(2) number of loads delivered; and (3) the average volumetric capacity of the vehicle container. The average volumetric capacity is obtained by measuring the volume of the vehicle chamber or container from the manufacturer's specifications and averaging the data for a random sampling of vehicles. The average delivered density of waste is determined by averaging the quotients of net load weight and the volumetric capacity of the vehicle container for a random sampling of vehicle loads. Total weight is the product of all three, namely, average density times average volume per load times number of loads. As is the case for the second method, the accuracy of the third method can be improved if data are collected by waste source. The third method is useful in particular for packer trucks. As is the case for the second method, the accuracy of the third method usually can be improved if the data are collected separately for residential, commercial, industrial, and self-haul sources.

Regardless of the method selected for quantity determinations, weight surveys should be conducted over a period of at least 1 week and initially of sufficient frequency to capture seasonal variations (e.g., two to four times per year). Thereafter the frequency over subsequent 12-month periods should be that necessary for meeting the desired level of accuracy, which is a function of the intended use of the data. Waste quantity studies that are conducted simultaneously with waste composition studies (to be described in detail subsequently) have two important benefits: (1) economies can be realized in labor and in costs due to the overlap of the planning and field work activities, and (2) the accuracy of the estimation of waste component quantities are greater than if the studies are conducted at different times. The accuracy of the waste component estimations is improved additionally if the data are collected separately for residential, commercial, industrial, and self-haul waste sources, or for specific subclasses of generators (e.g., single-family residences, office buildings, schools, etc.).

### *Rate of Waste Generation*

As described in the introduction of this chapter, current waste quantities and characteristics must be projected into the future in order to manage the waste over the long term. Thus, presumably accurately determined quantities and composition under the current conditions of the community must be estimated for conditions in the future.

Presently the accepted method of projection is to determine the current rates of generation on a unit basis and to multiply the unit quantities by the number of units estimated for convenient intervals of the planning period. Usually the planning horizons are 10, 15, or 20 years and the intervals are 1 or 5 years. Depending on the level of accuracy or of sophistication that is desired in the projections, the unit quantities can be based on actual and projected populations for projecting quantities for the entire waste stream (i.e., residential, commercial, industrial, and self-haul) or for residential generation, or they can be based on other relevant characteristics, e.g., number of employees or dollar of sales for commercial waste, etc. Examples of unit parameters and their values are illustrated in Table 3.6.

The procedure for projecting the current rate of generation to that of the future consists of the following:

1. Divide the total current quantity of waste or of a waste component by the unit parameter, e.g., population, thus producing a unit rate of generation. For illustration, in the case of residential waste the unit quantity could be lb/(capita · day).
2. Multiply the current unit rate of generation by the projected number of the unit parameter for the specified time interval, e.g., the future population to compute future lb/day.

### *Promotion of Cooperation*

A full explanation of the sampling program and its underlying rationale is an effective approach to enlisting the cooperation of the participants. Trained personnel in solid waste management are the individuals best qualified for estimating future waste quantities and composition from those currently measured. The data thus obtained are usually sufficient for arriving at an estimate of future waste generation sufficiently accurate (i.e., realistic) to meet most needs. Nevertheless, for completeness the numerical estimate should be verified quantitatively on a regular basis and even cross-checked with data from cities with similar patterns of growth, demographics, sources and generators of waste, waste quantities and composition, and climatical conditions.

**Table 3.6 Waste Generation Rates by Generator**

| Waste Generator Segment | September 1988 | December 1988 | March 1989 | June 1989 | Average | Units |
|---|---|---|---|---|---|---|
| Single-family residential[a] | | | | | | |
| Region 1 | 2.65 | 2.79 | 2.70 | 2.77 | 2.73 | lb/person/day |
| Region 2 | 2.69 | 2.40 | 2.53 | 2.55 | 2.58 | lb/person/day |
| Region 3 | 2.71 | 2.60 | 2.57 | 2.71 | 2.65 | lb/person/day |
| Average | 2.68 | 2.67 | 2.65 | 2.73 | 2.68 | lb/person/day |
| Apartments[b] | 1.7 | 2.3 | 3.36 | 2.4 | 2.5 | lb/person/day |
| Offices | 2.3 | 0.8 | 4.0 | 2.5 | 2.4 | lb/employee/day[c] |
| Eating and drinking establishments | 13.2 | 12.7 | 14.9 | 18.8 | 14.9 | lb/employee/day[c] |
| Wholesale and retail trade[d] | 0.021 | 0.036 | 0.018 | 0.007 | 0.021 | lb/$ sales |
| Food stores | 0.025 | 0.045 | 0.031 | 0.033 | 0.034 | lb/$ sales |
| Educational facilities | 0.5 | 0.6 | 0.4 | 0.5 | 0.5 | lb/student/day |

*Source:* CalRecovery, Inc.[4]

[a] Measurements for individual routes composed nearly entirely of single-family homes. Made over two consecutive weeks of collection during each month listed.
[b] Defined as three or more units
[c] Full-time equivalent.
[d] Except food stores.

## Composition

### *Sampling*

The generation of reasonably accurate data on composition of disposed waste demands an analysis period of at least 1 week, repeated seasonally, i.e., usually two to four times per year. During each of the seasonal sampling periods, samples to be analyzed should be taken from the collection vehicles and self-haul vehicles at a solid waste facility (e.g., transfer station, waste-to-energy facility, etc.). Each source of municipal solid waste (i.e., residential, commercial, industrial, and self-haul) should be sampled. The number of samples of each source of waste that is required for accurate representation of the parent waste source is a function of the variability to be expected in waste composition and the degree of accuracy sought for the determination of composition.

The results of many waste composition studies in the U.S. have shown that the variability in waste composition can be traced to the individual components and is typically a function of the sources of the waste. Some general trends are at least apparent in the U.S.:

1. Where extremes exist in growing conditions (i.e., disparate seasonal conditions) the variability in the percentage composition of yard waste among the seasons can be expected to be relatively large.
2. Components that comprise relatively low percentages in the waste stream (e.g., textiles, rubber, nonferrous metals, and household hazardous waste) tend to have large variability in percentage composition.
3. Taken as a whole, the variability among all components is generally lowest for residential waste and greatest for industrial wastes. The variability of commercial waste typically falls between that of residential waste and industrial solid waste.

One of the consequences of the difference in variability among the waste sources is that for the same level of accuracy typically more samples of industrial waste are required than for residential waste. However, conditions at a specific location may not be typical.

The statistical parameter that quantifies variability of measurements is the sample standard deviation. Thus, prior to the conduct of a waste composition sampling program, the standard deviations for the waste components of interest require estimation or must be inferred from an appropriately conducted study. Elementary statistics textbooks describe the procedures in both cases. An estimation of the average percentage composition of the components of interest is also necessary for the determination of number of samples. While not intended as a substitute for site-specific data, some typical "average" values of percentage composition and sample standard deviation are provided in Table 3.7 for use in the estimating process.

Lastly, an acceptable level of accuracy for the composition measurements must be established. Accuracy denotes the closeness of the average of a sample

**Table 3.7 Values of Mean ($\bar{x}$) and of Sample Standard Deviation (s) for Within-Week Sampling to Determine MSW Component Composition[a]**

| Component | Standard Deviation(s) | Mean ($\bar{x}$) |
|---|---|---|
| Mixed paper | 0.05 | 0.22 |
| Newsprint | 0.07 | 0.10 |
| Corrugated | 0.06 | 0.14 |
| Plastic | 0.03 | 0.09 |
| Yard waste | 0.14 | 0.04 |
| Food waste | 0.03 | 0.10 |
| Wood | 0.06 | 0.06 |
| Other organics | 0.06 | 0.05 |
| Ferrous | 0.03 | 0.05 |
| Aluminum | 0.004 | 0.01 |
| Glass | 0.05 | 0.08 |
| Other inorganics | 0.03 | 0.06 |
| Total | | 1.00 |

*Source:* CalRecovery.

[a] The tabulated mean values and standard deviations are estimates based on field test data reported for MSW from all waste sources sampled during weekly sampling periods at several locations around the U.S.

of measurements (the sample mean) to the average for the entire population (the population or true mean for the entire waste stream under consideration). The judgmental criteria are several:

1. What is the level of risk associated with the intended use of the data, e.g., is the data for use in general solid waste planning activities or for design of a facility? The greater the risk, the greater the level of accuracy that is required for the determinations.
2. What is the extent of funding that is available for the study?
3. Since the variability in waste composition fluctuates among the waste components and the computation of the level of accuracy is a function of variability (i.e., the standard deviation), a particular component must be selected as the limiting component. The limiting component determines the minimum number of samples for a specified level of accuracy.

In statistical parlance, measurement error (usually denoted as a percentage or decimal fraction) is the difference between the sample mean and the population mean.

The allowable degree of error in percent is equal to 100% minus the desired level of accuracy expressed in percent.

### *Sample Weight*

The minimum weight per sample should be on the order of 200 to 300 lb unless inordinately large particles comprise a substantial portion of the parent waste. The use of sample weights that are too small diminishes the likelihood of obtaining truly representative samples and therefore representative data. The increase in accuracy obtained with sample sizes larger than 200 to 300 lb normally does not warrant the additional effort required.

Beginning the analysis of the samples immediately after collection lowers the magnitude of error due to change in moisture content and due to any loss attributable to decomposition.

Figure 3.2 is a model of the type of data sheet suggested for the composition survey. This particular data sheet was developed in order to determine the viability of implementing a comprehensive recycling program. The model, therefore, accommodates a myriad of waste components. Time and cost to conduct a composition survey can be reduced by combining some categories. For example, mixed paper, newspaper, and cardboard may be combined under the single heading "paper" if such aggregating of waste components does not compromise the intent of the study. The analysis begins by sorting the wastes in the sample according to the categories listed on the data sheet. The sorting is done by placing each type of waste in its appropriate container. At the completion of the sorting, each container and its contents are weighed (gross weight). Gross weight and tare weight (i.e., empty container) should be recorded. The difference between gross weight and tare weight is the net weight of the content of the container. The usual presentation of the results is in the form of weight percent on an as-received (i.e., wet weight) basis. An example of results obtained in a comprehensive characterization of disposed waste is presented in Table 3.8. In many cases, modern waste management demands the identification and characterization of wastes generated by specific classes of generators in the community rather than a more general characterization of mixed MSW as a whole. Depending upon the conditions in the municipality and the needs of the waste management planners, these sectors may include multifamily dwellings, hotels, electronic manufacturers, colleges and universities (educational institutions), etc. The results of a waste composition study with the aforementioned level of sophistication was presented previously in Table 3.5.

## Cost of Quantity and Composition Studies

A 1-week study of waste quantities and composition typically costs in the range of $20,000 to $40,000. The range reflects in part the variation in the number of locations where sorting will occur, the desired extent of the breakdown of waste sources and of waste components, travel expenses, and reporting requirements.

## Other Physical Characteristics

Other physical characteristics to be determined can include moisture content, bulk density, and particle size distribution.

**WASTE COMPOSITION DATA SHEET FOR**

______________________________

JURISDICTION: ______________________
Day/Date: ______________________ Truck Co./No.: ______________________
Residential Sample No.: ______________ Truck Type: ______________________
Recorded By: ______________________ Waste Type: ______________________

| Component | Container Tare | Number of Tares | Gross Weights (lb) | | | | | | | Total |
|---|---|---|---|---|---|---|---|---|---|---|
| **Paper** | | | | | | | | | | |
| Corrugated containers | | | | | | | | | | |
| Mixed paper | | | | | | | | | | |
| Newspaper | | | | | | | | | | |
| High grade ledger paper | | | | | | | | | | |
| Other paper | | | | | | | | | | |
| **Plastics** | | | | | | | | | | |
| HOPE containers | | | | | | | | | | |
| PET containers | | | | | | | | | | |
| Film plastics | | | | | | | | | | |
| Other plastics | | | | | | | | | | |
| **Glass** | | | | | | | | | | |
| Refillable beverage | | | | | | | | | | |
| CA Redemption Value | | | | | | | | | | |
| Other recyclable | | | | | | | | | | |
| Other non-recyclable | | | | | | | | | | |
| **Metal** | | | | | | | | | | |
| Aluminum cans | | | | | | | | | | |
| Bi-metal containers | | | | | | | | | | |
| Ferrous metals/tin cans | | | | | | | | | | |
| Non-ferrous/aluminum scrap | | | | | | | | | | |
| White goods | | | | | | | | | | |
| **Yard waste** | | | | | | | | | | |
| **Other organics** | | | | | | | | | | |
| Food wastes | | | | | | | | | | |
| Tires/rubber products | | | | | | | | | | |
| Wood wastes | | | | | | | | | | |
| Agricultral crop residues | | | | | | | | | | |
| Manure | | | | | | | | | | |
| Textiles/leather | | | | | | | | | | |
| Miscellaneous | | | | | | | | | | |
| **Other wastes** | | | | | | | | | | |
| Inert solids | | | | | | | | | | |
| HHW | | | | | | | | | | |
| | | | | | | | | | | |

**Figure 3.2.**

**Table 3.8 Results of Comprehensive Waste Characterization Program**

| | Fall 1988 | Winter 1988 | Spring 1989 | Summer 1989 | Average[a] |
|---|---|---|---|---|---|
| Organics | 86.0 | 86.5 | 87.7 | 89.8 | 87.5 |
| Paper | 44.7 | 45.7 | 47.5 | 40.3 | 44.5 |
| OCC/kraft-recyclable | 2.6 | 3.8 | 2.7 | 2.4 | 2.9 |
| OCC/kraft-contaminated | 8.1 | 3.7 | 5.8 | 4.8 | 5.6 |
| Newsprint | 9.4 | 11.0 | 8.2 | 8.1 | 9.2 |
| High grade | 3.7 | 6.5 | 6.6 | 6.1 | 5.7 |
| Magazines | 1.6 | 1.8 | 2.6 | 2.5 | 2.1 |
| Mixed paper | 16.9 | 12.8 | 15.1 | 7.9 | 13.2 |
| Other paper | 2.2 | 6.1 | 6.7 | 8.5 | 5.9 |
| Plastic | 6.1 | 6.5 | 7.0 | 6.0 | 6.4 |
| Film | 3.6 | 3.4 | 3.8 | 3.1 | 3.4 |
| HDPE | 0.4 | 0.3 | 0.5 | 0.4 | 0.4 |
| PET | 0.1 | 0.1 | 0.1 | 0.1 | 0.1 |
| Polystyrene foam | 0.4 | 0.6 | 0.6 | 0.3 | 0.5 |
| Other plastic | 1.6 | 2.0 | 2.0 | 2.1 | 1.9 |
| Yard waste | 15.0 | 15.1 | 7.4 | 15.0 | 13.1 |
| Firewood | 0.0 | 0.3 | 0.0 | 0.0 | 0.1 |
| Other plant matter | 15.0 | 14.9 | 7.4 | 15.0 | 13.1 |
| Wood | 1.4 | 0.8 | 1.0 | 0.8 | 1.0 |
| Food | 15.2 | 14.3 | 18.0 | 21.5 | 17.3 |
| Textiles | 1.5 | 1.3 | 1.3 | 2.0 | 1.5 |
| Other organics | 2.3 | 2.9 | 5.4 | 4.2 | 3.7 |
| Inorganics | 13.4 | 12.8 | 12.2 | 9.8 | 12.1 |
| Metals | 3.0 | 3.9 | 4.0 | 3.1 | 3.5 |
| Steel food cans | 1.7 | 1.9 | 2.1 | 1.6 | 1.8 |
| Other ferrous | 0.7 | 1.1 | 1.3 | 0.9 | 1.0 |
| Aluminum cans | 0.5 | 0.4 | 0.4 | 0.3 | 0.4 |
| Other aluminum | 0.2 | 0.3 | 0.2 | 0.2 | 0.2 |
| Other metals | 0.0 | 0.2 | 0.0 | 0.1 | 0.1 |
| Glass | 7.0 | 7.1 | 7.1 | 5.6 | 6.7 |
| Redeemable glass | 4.3 | 3.3 | 2.5 | 2.4 | 3.1 |
| Wine & liquor | 1.1 | 1.6 | 1.5 | 1.9 | 1.5 |
| Other container glass | 1.3 | 2.0 | 2.9 | 1.2 | 1.9 |
| Other glass | 0.3 | 0.3 | 0.2 | 0.2 | 0.2 |
| Soil | 0.9 | 0.4 | 0.0 | 0.0 | 0.3 |
| Other inorganics | 2.5 | 1.4 | 1.1 | 1.2 | 1.5 |
| Special wastes | 0.5 | 0.6 | 0.1 | 0.3 | 0.4 |
| Appliances | 0.2 | 0.3 | 0.0 | 0.2 | 0.2 |
| Chemicals | 0.1 | 0.2 | 0.1 | 0..1 | 0.1 |
| Reusable | 0.3 | 0.1 | 0.0 | 0.0 | 0.1 |
| Total | 100 | 100 | 100 | 100 | 100 |
| Vehicles Sampled | 16 | 14 | 16 | 14 | |

*Source:* CalRecovery, Inc.[4]

[a] Averages may not sum to 100.0% due to rounding.

### Air-Dry Moisture Content

The procedure for determining air-dry moisture content is as follows:

1. Weigh the sample as received (''wet weight'').
2. Air-dry until the moisture content of the sample is in equilibrium with that of the ambient air.
3. Determine the moisture content through the following formula:

$$\text{Moisture content (\%)} = \frac{W_w - W_o}{W_w} \times 100$$

where $W_w$ = wet weight of sample and $W_o$ = dry weight of sample.

### Bulk Density

Inasmuch as they affect bulk density, the identity of the type of material and degree of preprocessing should be determined as part of the design of material recovery systems. For example, crushed glass has a higher bulk density than do unprocessed glass containers. Not unexpectedly, raw waste contained in a compactor collection truck has a higher density than it would have if contained in a conventional dump truck. Bulk density also is affected by moisture content. Bulk density of the waste determines the space requirements for storage and has a significant bearing on equipment specifications.

Bulk density of a feedstock before processing (e.g., raw waste), during processing, and after it has been processed can be measured by filling a container of known volume with the material and weighing the loaded container (the container should be shaken during filling to reduce voids). The bulk density is calculated by dividing the net weight of the material (weight of loaded container minus weight of empty container) by its volume. The result is expressed as pounds per cubic feet. Average densities for mixed MSW and for various components of the waste stream are presented in Table 3.9.

### Size Distribution of Unprocessed MSW

If circumstances indicate a need to determine the particle size distribution of raw MSW (e.g., to estimate material recovery by trommel screening), it can be done with the use of a set of at least four manually mechanically agitated screens. The screens should have square openings if the standard screening precedents are of concern. Size of openings in the first screen should be 8 in.; in the second screen, 4 in.; in the third screen, 2 in.; and in the fourth, 1 in. The weight of the sample should be within the range of 100 to 150 lb. Multiple batches will likely be required to avoid overloading the screens. The procedure is as follows:

**Table 3.9 Average Densities of Mixed MSW and Components Derived Therefrom**

| Component | Density ($lb/ft^3$) |
|---|---|
| Mixed MSW | |
| Loose | 5.6–11.1 |
| After dumping from compactor truck | 12.9–14.8 |
| In compactor truck | 18.5–25.9 |
| In landfill | 29.6–48.1 |
| Shredded | 7.4–14.8 |
| Baled | 29.6–44.4 |
| Loose bulk densities | |
| Corrugated | 1–2 |
| Aluminum cans | 2–3 |
| Plastic containers | 2–3 |
| Miscellaneous paper | 3–4 |
| Garden waste | 4–5 |
| Newspaper | 5–7 |
| Rubber | 13–16 |
| Glass bottles | 15–19 |
| Food waste | 22–25 |
| Tin cans | 4–5 |
| True densities | |
| Wood | 37 |
| Cardboard | 43 |
| Paper | 44–72 |
| Glass | 156 |
| Aluminum | 168 |
| Steel | 490 |
| Polypropylene | 56 |
| Polyethylene | 59 |
| Polystyrene | 65 |
| ABS | 64 |
| Acrylic | 74 |
| Polyvinylchloride (PVC) | 78 |
| Material fractions | |
| Mixed MSW processing facilities | |
| dRFD | 30–40 |
| Aluminum scrap | 14–16 |
| Ferrous scrap | 23–26 |
| Crushed glass | 65–85 |
| Powdered RDF (Eco-Fuel) | 26–28 |
| Densified materials | |
| Baled aluminum cans | 12–18 |
| Cubed ferrous cans | 65–93 |
| Baled corrugated | 22–32 |
| Baled newspaper | 23–33 |
| Baled high grades | 20–29 |
| Baled PET | 13–19 |
| Baled HDPE | 17–24 |

*Source:* CalRecovery, Inc.[8]

**SIZE DISTRIBUTION DATA SHEET**

Date: ______________________ Sample Wet Weight: ______________________
Site: ______________________ Sample Dry Weight: ______________________
Test No.: ______________________ Water Content: ______________________
Grinder: ______________________ Moisture Content: ______________________
Material: ______________________ Screening Time: ______________________

| Screen Size | Bottom Screen Size ( ) | Gross Wt. Retained by Bottom Screen ( ) | Tare Wt. ( ) | Net Wt. Retained by Bottom Screen ( ) | % of Feed on Bottom Screen | Cumulative Wt. % Passing Bottom Screen |
|---|---|---|---|---|---|---|
| | | | | | | |
| | | | | | | |
| | | | | | | |
| | | | | | | |
| | | | | | | |
| | | | | | | |
| | | | | | | |
| | | | | | | |
| | | | | | | |
| | | | | | | |
| | | | | | | |
| | | | | | | |
| | | | | | | |
| Pan | | | | | | |
| | | Total Sample Wt. | | | | |

NOTES: ______________________

**Figure 3.3.**

1. Place the sample on the screen having the 8-in. openings. Shake the screen until particles no longer pass through the openings. Do not overload the screen such that particles cannot find their way to the screen surface during the process of agitation. Material retained on the screen (''oversize'') is collected and weighed.
2. Material that passed through the screen (''undersize'') is placed on the screen with the 4-in. openings. Shake the screen and proceed as in Step 1.
3. Repeat the process until all four screens have been used.
4. Fractions that are sized are weighed, and the weight values are used to plot a size distribution curve. Typically, the size distribution is plotted as cumulative weight percent passing through vs screen size. A sample data sheet is shown in Figure 3.3 and a sample size distribution curve is shown in Figure 3.4.

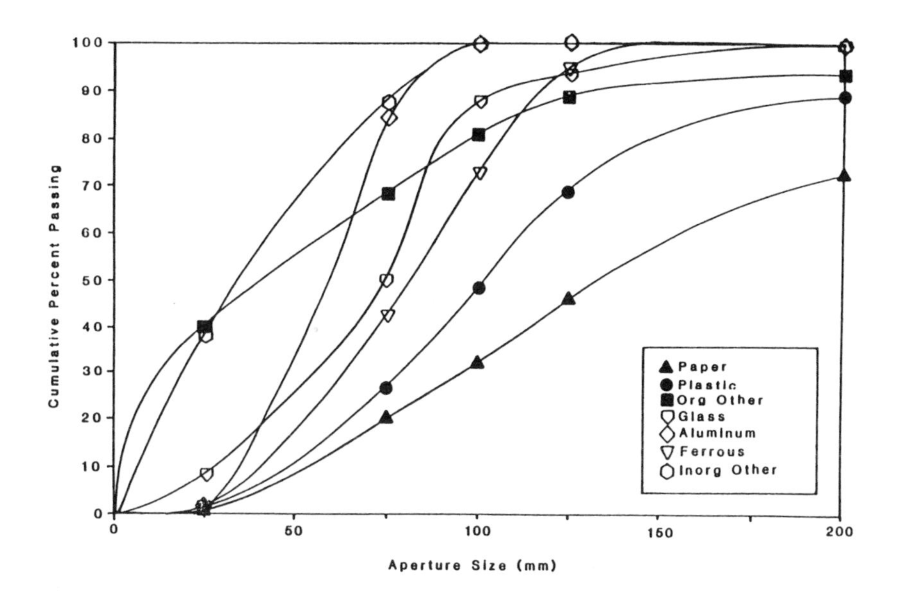


**Figure 3.4.**

## STANDARD TEST METHODS

Waste characterization is one of the very few areas of solid waste management in which standardized methods of testing and analyses have been promulgated. For purposes of reference some methods that are applicable or adaptable to solid wastes and fractions or materials derived therefrom are summarized in Table 3.10.

## SUMMARY

The sampling of waste is a necessary step in the determination of the generation rate, quantity, composition, and other relevant properties of the wastes to be collected and processed. A satisfactory knowledge of the these characteristics is a prerequisite for rational planning and designing of any waste management undertaking. This is particularly true when composting and recycling are to be key elements of the undertaking. The reason is that the characteristics exert a serious impact on the design and operational characteristics of such an undertaking. Hence, it is vital that quantity, composition, and applicable properties over time be accurately determined.

Waste quantities and characteristics influence several critical aspects of an operation, such as volume, number and frequency of vehicles serving the operation, land area required, and yield and quality of end products, as well as other ancillary requirements.

**Table 3.10 Concensus Standard Methods Applicable or Adaptable to Characterization of Solids Waste or Fractions Derived Therefrom**

| Number ASTM[a] | Title |
|---|---|
| Methods for | |
| E 954-88 | Packaging and Shipping of Laboratory Samples for Refuse-Derived Fuel |
| E 829-89 | RDF Laboratory Samples for Analysis, Preparing |
| E 791-87 | Refuse-Derived Fuel Analysis Data from As-Determined to Different Bases, Calculating |
| E 926-88 | Refuse-Derived Fuel (RDF) Samples for Analyses of Metals, Preparing |
| E 828-81 (1987)[1] | Size of RDF-3 from Its Sieve Analysis, Designating |
| Test methods for | |
| E 885-88 | Analyses of Metals in Refuse-Derived Fuel by Atomic Absorption Spectroscopy |
| E 830-87 | Ash in the Analysis Sample of Refuse-Derived Fuel |
| E 777-87 | Carbon and Hydrogen in the Analysis Sample of Refuse-Derived Fuel |
| E 889-82 (1988) | Composition or Purity of a Solid Waste Materials Stream |
| E 1109-86 | Determining the Bulk Density of Solid Waste Fractions |
| E 886-88 | Dissolution of Refuse-Derived Fuel (RDF) Ash Samples for Analyses of Metals |
| E 776-87 | Forms of Chlorine in Refuse-Derived Fuel |
| E 953-88 | Fusibility of Refuse-Derived Fuel (RDF) Ash |
| E 711-87 | Gross Calorific Value of Refuse-Derived Fuel by the Bomb Calorimeter |
| E 778-87 | Nitrogen in the Analysis Sample of Refuse-Derived Fuel |
| E 790-87 | Residual Moisture in a Refuse-Derived fuel Analysis Sample |
| E 955-88 | Thermal Characteristics of Refuse-Derived Fuel Macrosamples |
| E 775-87 | Total Sulfur in the Analysis Sample of Refuse-Derived Fuel |
| Methods of testing | |
| E 701-80 | Municipal Ferrous Scrap |
| E 887-88 | Silica in Refuse-Derived Fuel (RDF) and RDF Ash |
| E 897-88 | Volatile Matter in the Analysis Sample of Refuse-Derived Fuel |

[a] American Society for Testing and Materials.

## REFERENCES

1. Diaz, L.F. and C.G. Golueke, "Solid Waste Management in Developing Countries", *BioCycle*, September (1985).
2. Diaz, L.F. and C.G. Golueke, "Solid Waste Management in Developing Countries", *BioCycle*, July (1987).
3. U.S. Environmental Protection Agency, "Design, Construction, and Evaluation of Clay Liners for Hazardous Waste Facilities", EPA/530-SW-86-007, PB86-18496/AS, NTTS, Springfield, VA.
4. CalRecovery, Inc., "Waste Characterization Study for Berkeley, California", prepared for the Department of Public Works, City of Berkeley, CA, December (1989).
5. Diaz, L.F., G.M. Savage, and C.G. Golueke, *Resource Recovery from Municipal Solid Wastes* (Boca Raton, FL: CRC Press, 1982).
6. Japan International Cooperation Agency, "Master Plan and Feasibility Study on Seoul Municipal Solid Waste Management System", Tokyo, Japan, 1985.

7. CalRecovery, Inc., ''Broward County Resource Recovery Report — Waste Characterization Study'', prepared for Resource Recovery Office, Broward County, Florida, February (1988).
8. CalRecovery, Inc., ''Conversion Factor Study — In-Vehicle and In-Place Waste Densities'', prepared for the California Integrated Waste Management Board, March (1992).

# CHAPTER 4

# Processing

## INTRODUCTION

Chapters 4 and 5 expand upon the coverage of the other chapters in that they deal with the entire MSW stream, i.e., the nonbiodegradable and the biodegradable components. The processing about which Chapter 4 is concerned takes place directly after the wastes have been delivered to a treatment facility or a treatment/final disposal facility. If the treatment facility is an entity separate from the disposal facility, rejects from the treatment facility are disposed at the final disposal facility. In other cases, some processing may be done at a transfer station and the remainder is done at a treatment or treatment/disposal facility.

In practice, source separation will never be so thorough that processing would not be a necessary part of the separation and recovery of any one or more of the recyclable resources in MSW. This is true regardless of whether the desired resource is organic matter for composting, useful metals, glass, or any other component. Processing begins with the separation (recovery) of the resource from the other components of the waste stream and the readying of the separated resources for treatment and ends with the rendering of them suitable for their destined uses. Although certain compost systems have been proposed and implemented which call for no processing of the incoming MSW, other than perhaps some particle size reduction, the general experience with such systems has been much less than satisfactory. This has been true both with respect to the composting process itself and to the product.

Processing for resource recycling involves a series of unit processes, the number of which depends upon that of the types of resources to be recycled and the extent to which they have been source separated. The nature and design of each unit process are those which accommodate the physical and chemical characteristics of the particular resource or category of resources for which it is intended. Because most categories of resources have certain characteristics in common, basic processing principles (e.g., those pertaining to size reduction, air classification, and screening) usually are nonspecific with respect to the resource or resources in a category, although details of equipment design, the size, degree of complexity, and the cost of an individual unit process are strongly influenced by the nature and utility of the resource to be recovered and the extent and degree of source separation. The gradation also depends upon the function being served, e.g., separation and recovery from the waste stream, readying the recovered resource for reuse.

Whereas this chapter (Chapter 4) deals with the generalities of recycling, Chapter 5 is concerned with the details. Accordingly, basic principles of unit process categories and types of equipment pertaining to them are described and discussed in general terms in this chapter. Fitting the unit processes to specific operations to attain particular goals is the subject in Chapter 5. Thus, arrangement and sequence of unit processes (flow diagrams) as applied to MRFs are reserved for Chapter 5. Furthermore, because Chapter 4 concerns generalities, it applies to mixed MSW. The extent and nature of modifications to the design for source separated MSW also are reserved for Chapter 5.

## SEPARATION (Recovery)

Of the categories of unit processes involved, those related to physical separation and removal from the waste stream are the most essential; if for no reason other than logic, separation is the first step. The separation unit process differs from the other unit processes in that it can be done either manually or mechanically, or with a combination of the two.

### Manual Separation

Bulky items (appliances, furniture, etc.) and other specified items (e.g., hazardous waste) are manually removed from the waste prior to processing. With few exceptions, a completely manual separation beyond this initial separation is reserved for small operations less than 20 TPD. However, it also may be applicable to the removal of "contaminants" from separately collected resources. (Here, "contaminants" refers to components other than the desired resource.)

Equipment involved in manual removal usually is a sorting belt or table on which the waste moves past workers ("sorters") stationed on one or both sides of the belt or table. Hoppers or other receptacles for receiving removed items are positioned within easy reach of the sorters.

The design of processes which rely on manual separation require a good understanding of basic principles dealing with time and motion as well as comfort and safety of the sorters. Thus far, these considerations have been grossly overlooked in several facilities.

### Mechanical Separation

Mechanical separation involves the use of several types of unit processes, five of which are size reduction, screening, air classification, magnetic separation, and aluminum separation.[1,2,3,4,5] Table 4.1 lists most of the major equipment categories that have been incorporated into facilities implemented to date.[6]

**Table 4.1 Mechanical Equipment Used in Material Recovery Systems**

1. Size reduction/shredding
   a. Hammermills — vertical and horizontal shaft
   b. Shear shredder
   c. Rotary, guillotine, and scissors-type shears
   d. Grinders — roller, disc-mill, ball mill
   e. Flail mill
   f. Wet pulper
   g. Knife mill

2. Air classifiers
   a. Vertical straight
   b. Vertical zig-zag
   c. Vibrating inclined
   d. Horizontal straight
   e. Inclined rotating drum
   f. Density separators (stoners, etc.)
   g. Air knife

3. Screens
   a. Trommel
   b. Vibrating — reciprocating and gyrating
   c. Disc

4. Magnetic separator
   a. Belt type
   b. Drum type

5. Glass and aluminum separators
   a. Heavy media separation
   b. Aluminum magnets (eddy current separation)
   c. Froth flotation units
   d. Optical sorting

6. Dryers
   a. Drum type
   b. Fluid type

7. Densifiers
   a. Pelletizers
   b. Briquetters
   c. Cubers
   d. Extruders
   e. Compactors
   f. Balers
      (1) Rectangular prism shapes
      (2) Flat cylindrical shapes
   g. Can flatteners

8. Handling equipment
   a. Front-end loaders
   b. Grapples
   c. Conveyors — belt, apron, bucket elevator, pneumatic, drag
   d. Forklifts

The sequence of the processes varies, although either size reduction or a preliminary screening (trommel) is the first step. Flow patterns and their variations are described in Chapter 5. The intent of this section is to present a brief overview of the equipment; in-depth discussions can be found in Reference 1.

### *Size Reduction*

As used in solid waste management, the term "size reduction" has several synonyms, four of which are "shredding," "milling," "comminution," and "grinding." The term "shredding" has been widely adopted in reference to size-reducing refuse.

Size reduction is an essential step in mechanical separation because it facilitates handling and reduces bulky items to particles, the sizes of which are compatible with the processing equipment. Size reduction brings about a degree of uniformity in terms of the maximum particle size of the diverse components of the incoming waste stream. This uniformity is a requirement of most mechanical sorting systems.[1]

Coarse or primary shredding, grinding, or milling, usually to a minimum particle size of about 4 in., is a feature of practically all resource recovery facilities. Secondary and even tertiary shredding (collectively called "fine" shredding) are introduced whenever a particle size significantly smaller than 4 in. is specified (e.g., the production of a refuse-derived fuel of small particle size[4]). Other applications in which fine shredding may be involved are the recovery and processing of ferrous metals, aluminum, plastic, and glass scrap to meet user specifications.

The practice of size reduction in MSW management is not solely a resource recovery activity — it also can be used as a preparatory step to landfilling. Here the purpose is to improve the overall quality of a landfill as well as to increase landfill capacity.

*Types of Shredders.* Low-speed, high-torque flail mill-type shredders and shear shredders are also used for size reducing solid waste. However, the utilization generally is for coarse shredding. The hammermill is a type of high-speed shredder frequently used for size reducing solid waste.[3,7] It is an impact crusher in which a combination of tensile, compressive, and shear forces is applied to the throughput material by rotating elements (hammers) that strike particles of the material while they are in suspension, or that hurl the particles at a high speed against stationary surfaces.[1]

Hammermills. The hammers or cutters and the grate bars, if any, are the two components of a hammermill that are exposed to the most extensive and intensive wear. The reason is that these components come into direct and continuous contact with the throughput wastes. As a precautionary measure, a fire and

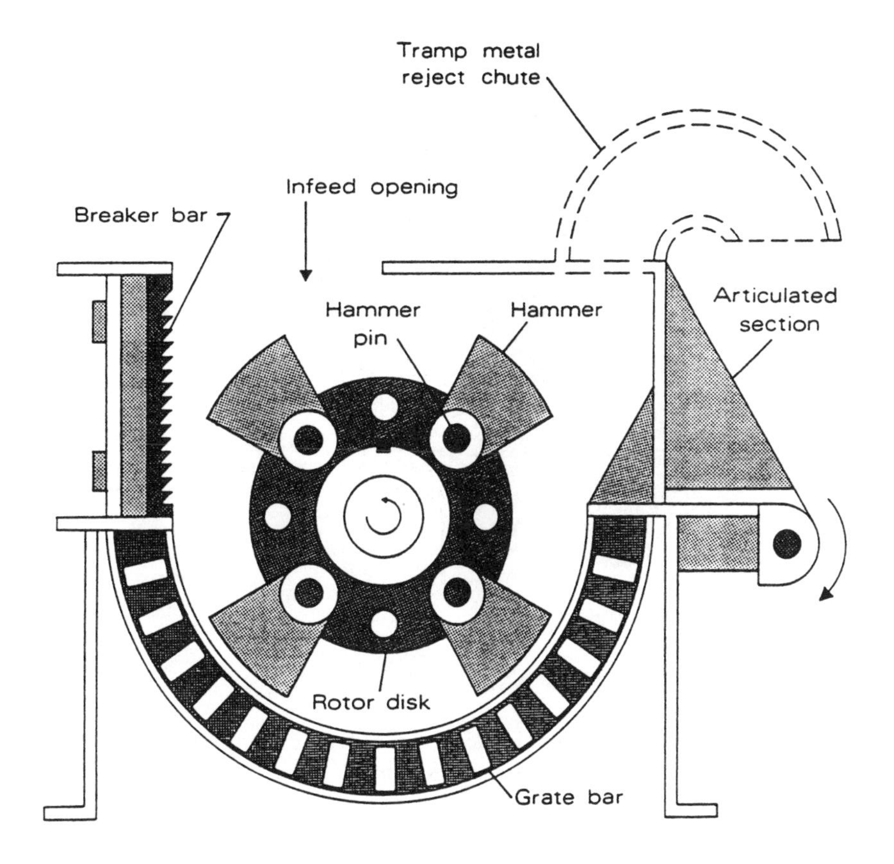


**Figure 4.1.** Horizontal swing hammermill (cross-sectional view).

explosion suppression system can and should be incorporated into the shredder design. This is particularly important when mixed solid waste is used.

Hammermills can be divided into two broad groups on the basis of orientation of the rotor, namely horizontal and vertical. Both groups have elements (i.e., hammers) that rotate within the shredder and effect particle size reduction through collision with the infeed material. The hammers may be rigidly attached to the shredder rotor or freely hinged (swing hammers). The horizontal swing hammer is the type most commonly used. Its principal parts are the rotor, hammers, grates, frame, and flywheel. Its rotor and flywheel are mounted through bearings to the frame. The bottom portion of the frame also holds the grates. Figure 4.1 is a diagrammatic sketch of a horizontal hammermill.

In a horizontal hammermill, the hammers extend perpendicularly to the rotor when the entire rotor-hammer assembly is rotating (usually at 1000 to 1500 RPM). Upon impact with an object being size reduced, a hammer is free to move in a 180° arc about the connecting pins. Objects to be size reduced are introduced into the machine by way of an infeed chute. They then interact with

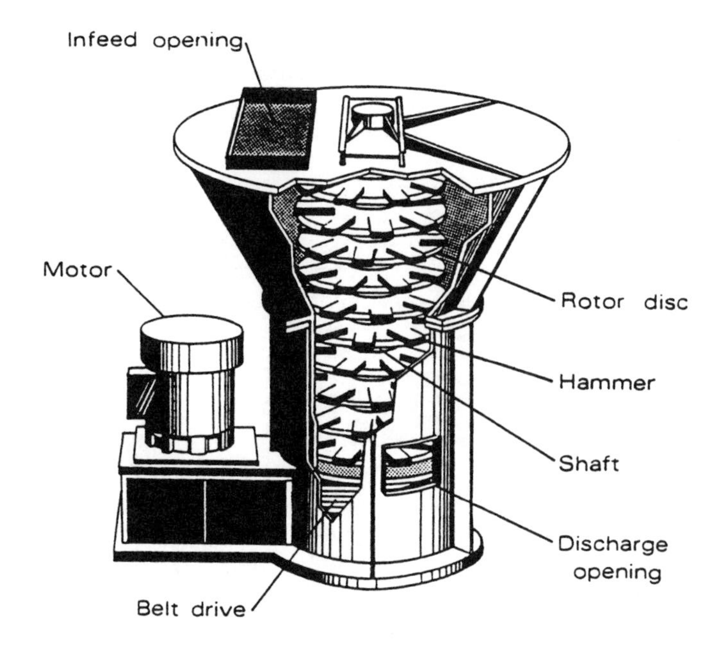


**Figure 4.2.** Vertical hammermill.

the hammers and each other until at least one of their dimensions reaches a size small enough for the particle to fall through the grates at the bottom of the machine.

Residence time of the material in the mill and the size distribution of the discharge (product) are largely determined by grate spacing. Other factors that affect product size distribution are feed rate, moisture content, and mill speed (i.e., velocity of the tip of the hammer).

A diagrammatic sketch of the vertical type of hammermill is presented in Figure 4.2. As is indicated in the figure, the rotor is placed in a vertical position. The input, assisted by gravity, drops parallel to the shaft axis and is exposed to the action of the hammers. It is shredded by the time it is discharged at the bottom of the machine.

Because both the horizontal and the vertical types of shredders are inherently somewhat self-destructive, their maintenance is an important consideration. To minimize downtime and labor, the following items should be incorporated into the overall shredder concept:

1. Appropriate equipment (e.g., overhead cranes, hydraulic or pneumatic devices for opening the shredder for access to the internal parts and for closing it, mechanical pin pullers, hammer removal jigs, and reversible hammers).
2. Appropriate and well-kept welding equipment for, primarily, resurfacing hammers and facilities that include a properly designed ventilation system.
3. The lighting required for the safe performance of maintenance operations.

4. Moveable scaffolding or permanent walkways about the shredder to permit easy access to all pieces of equipment that may require maintenance.
5. Adequate space around the shredder to accommodate the removal and reinstallation of grate bars, hammers, wear plates, breaker bars, and other parts.
6. Explosion and fire control systems.
7. Dust collectors.
8. Electrical equipment and controls.
9. An inventory of the manufacturer's recommended spare parts including those which are required for bearing and rotor replacement or overhaul, and for the removal and reinstallation of hammers, liners, wear plates, breaker bars, and grate bars or cages.

Shear Shredders. One of the most modern innovations in equipment used to size reduce MSW is the high-torque, low-RPM shredder. This shredder is sometimes known as a "shear" shredder. The design of the shear shredder is presented in Figure 4.3. As shown in the figure, the unit consists of two horizontal, counterrotating shafts. Each shaft contains cutters to tear and shear the material. In shear shredders, shear and deformation are the primary mechanisms of particle size reduction. The cutters usually operate within a range of 20 to 70 RPM. Because of the shearing action and high torque availability, shear shredders are used to size reduce items that are difficult to shred such as tires.

*Characteristics of Size Reduction Processes.* Size reduction of refuse and its components is an energy- and maintenance-intensive operation. Energy requirements for size reducing some solid waste fractions in one type of hammermill shredder are indicated in Figure 4.4. In the figure, specific energy (i.e., kilowatt hours per ton processed) is plotted as a function degree of size reduction. Degree of size reduction ($Z_o$) is defined as unity minus the ratio of characteristic product size ($X_o$) to the feedstock characteristic size ($F_o$).[3] The characteristic particle size is the screen size on a size distribution curve corresponding to 63.2% cumulative weight passing. The curves in the figure demonstrate that composition influences the energy requirements for size reduction. For example, screened light fraction (primarily paper and plastic) requires more energy on a unit basis to achieve a given degree of size reduction than does mixed MSW.

Energy requirements for the size reduction of MSW vary somewhat depending on the type and design of the size reduction equipment. The requirements for size reducing mixed MSW to various product sizes using different types of horizontal hammermills are illustrated in Figure 4.5. The curves shown in the figure are the results of the field testing of a variety of size reduction devices. Illustrated in the figure is the fact that the specific energy requirement increases substantially if product sizes less than 1 to 2 cm are required.

Hammer wear represents a substantial operating expense of size reduction of solid waste. However, wear and its associated costs can be controlled if the

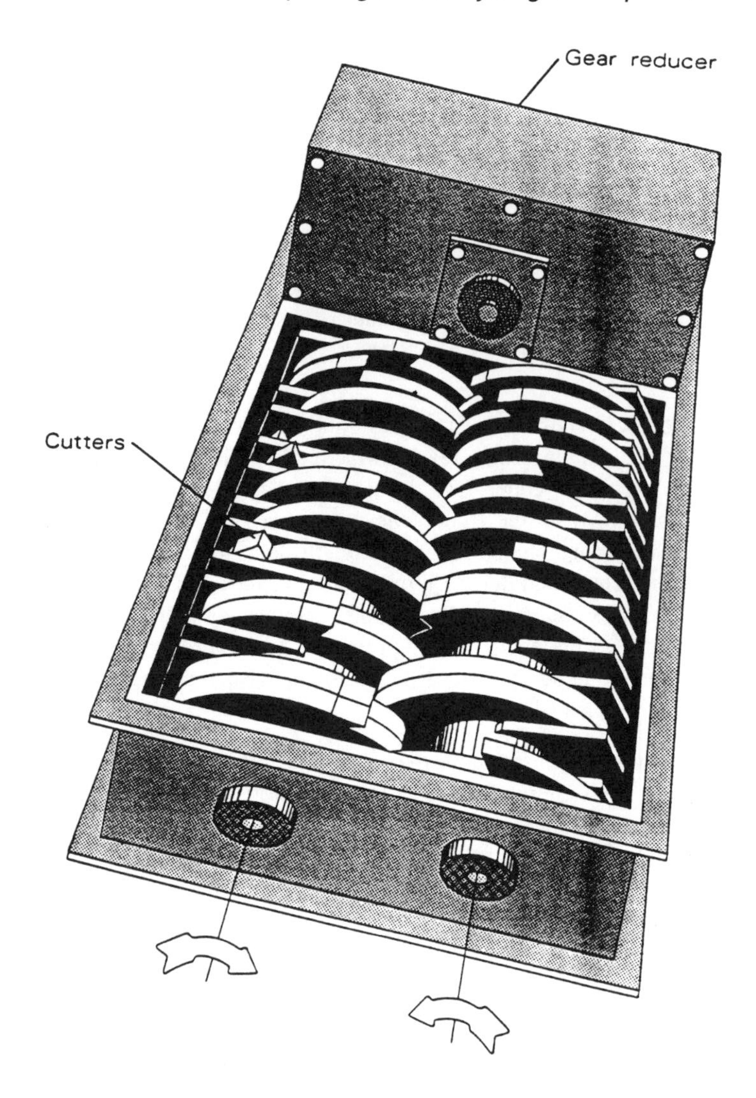


**Figure 4.3.** Low-RPM/shear-type shredder.

properties of the waste, operating characteristics of the size reduction device, and metallurgy of the hammers are taken into consideration. Selection of hammer parent metallurgy and hardfacing metallurgy based on properties of the feedstock and on operating conditions can minimize hammer wear. For example, the authors have shown that relatively hard surfacing materials (as characterized by the Rockwell C hardness scale) exhibit substantially less wear in the size reduction of MSW than softer (and often standardly supplied) commercial metallurgical formulations for parent hammer material and for hardfacings. The effect is illustrated in Figures 4.6 and 4.7 for the size reduction of mixed MSW using vertical and horizontal hammermills.

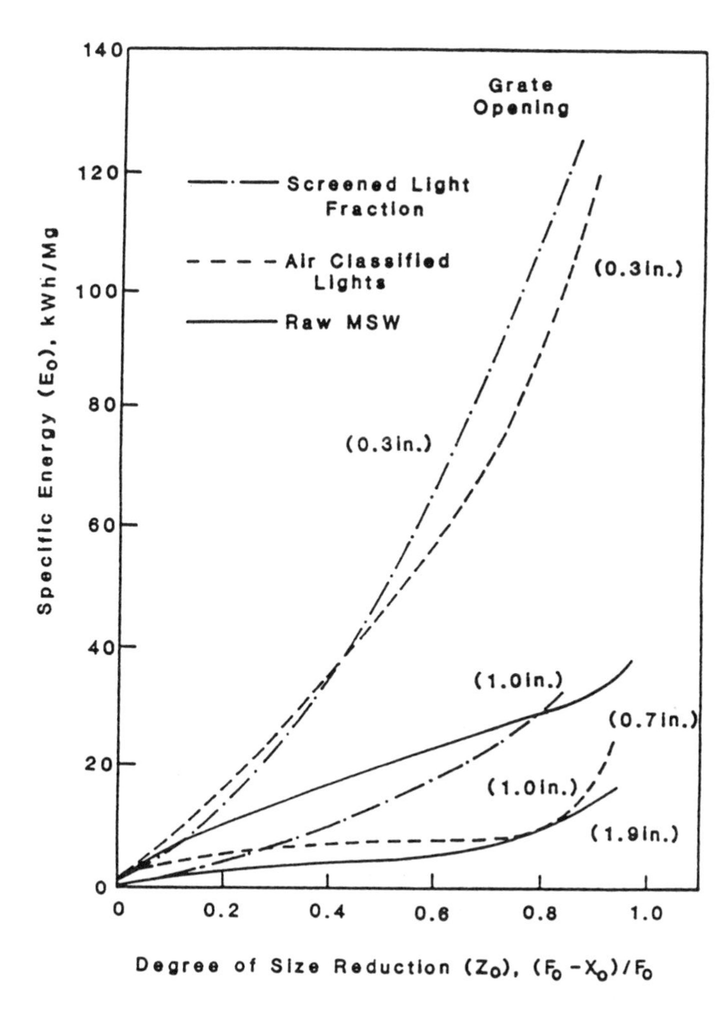


**Figure 4.4.** Energy requirements for selected solid waste fractions using a 187-kW horizontal swing hammermill.

## *Air Classification*

Air classification is a means of separating categories of materials by way of differences between their respective densities. The underlying mechanism is the interaction between a moving air stream and shredded waste material within a column. As a result of the interaction, drag forces are exerted upon the particles. However, the drag forces are simultaneously opposed by the gravitational force exerted on the waste particles. Consequently, waste particles that have a large drag-to-weight ratio are suspended in the air stream; whereas components that have a small ratio tend to settle out of the air stream. The suspended fraction conventionally is referred to as the "air-classified light fraction" and the settled

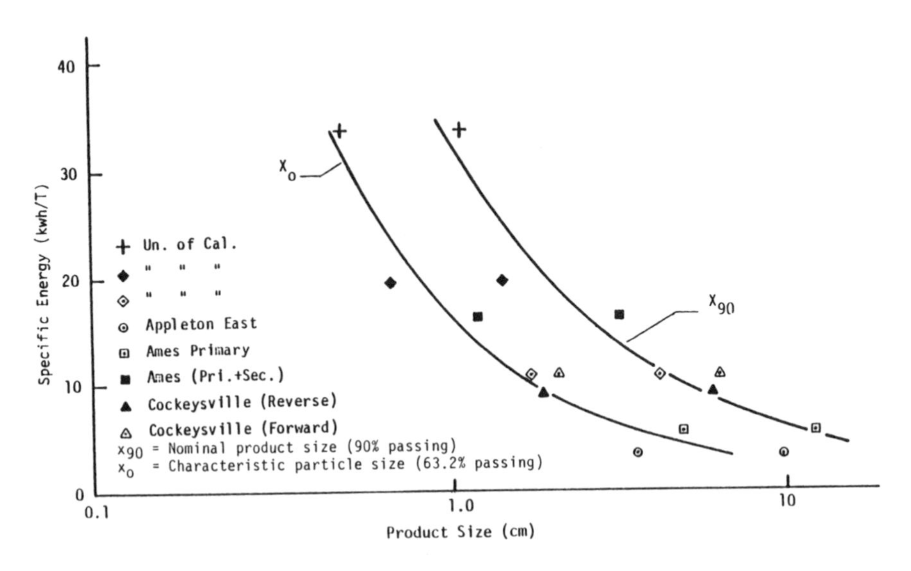


**Figure 4.5.** Specific energy requirement (wet weight basis) as a function of product size.

fraction is termed "air-classified heavies." The two fractions, respectively, may be referred to simply as the "lights" and the "heavies." The unit in which the separation takes place is designated an "air classifier."

In air classification of shredded, mixed MSW, paper and plastic tend to be concentrated in the light fraction, and metals and glass are the principal components of the heavy fraction.

Density is not the only variable that affects the air classification process. For example, by virtue of its high drag-to-weight ratio, fine glass may appear in the light fraction. On the other hand, flat, unshredded milk cartons may appear in the heavy fraction. In addition, moisture affects the separation of the various components, in particular the paper fraction, from one another.

Air classifiers may be one of a number of designs. The three principal groups of designs are diagrammed in Figures 4.8 to 4.11. All three require appurtenant dust collection, blower, separator, and conveying facilities.[1]

Air classification can have an advantageous effect on the output of other separation unit processes. Thus, the quality of the magnetically recovered ferrous fraction can be substantially improved by removal of residual paper and plastic in an air classifier. A version of air classifier can be used to clean up the mixed nonferrous material generated by eddy current processing in the removal of nonferrous metals. In the two applications, air classification serves to (1) remove light organic matter entrained with the metal; and (2) separate light aluminum from heavier aluminum castings, copper, bronze, etc.

## *Screening*

Screens are used for achieving efficient separation of refuse particles through reliance on differences between particle sizes with respect to any two dimensions.

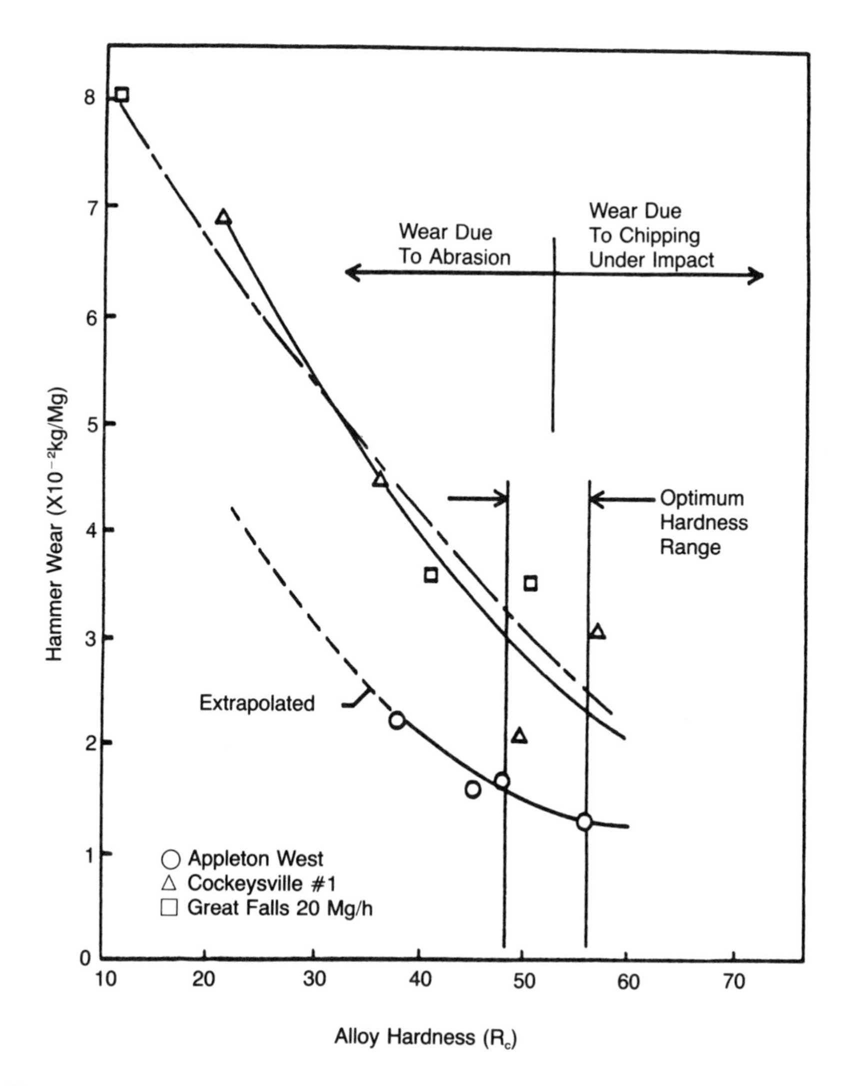


**Figure 4.6.** Hammer wear associated with MSW size reduction as a function of alloy hardness.

In the simplest case, the separation results in a division of the incoming material into two size groups, one of which has a minimum particle size larger than that of the individual screen openings; whereas the second has a maximum particle size smaller than that of the openings. The first group is retained on the screen. This group is termed "oversize," and its constituent particles become "oversized particles." The second group passes through the openings and accordingly is termed "undersize," and its constituent particles become "undersized particles."

Screens may also be used to separate feed stream into streams corresponding to three or more size classes. In such cases, several screen surfaces of different size openings are fitted in series in the frame of screening equipment.

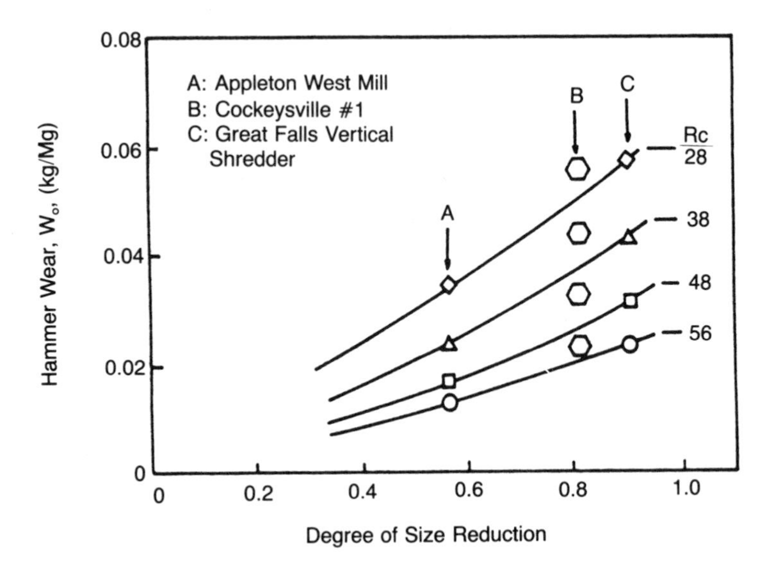


**Figure 4.7.** Hammer wear associated with MSW size reduction as a function of alloy hardness and degree of size reduction.

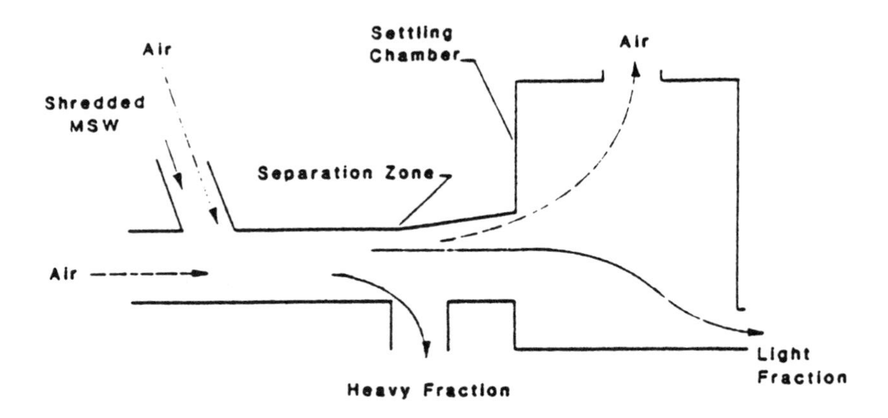


**Figure 4.8.** Horizontal air classifier.

Only three of the several types of screens employed in industry are used for sizing particular fractions of processed and unprocessed MSW. The three types are the vibratory flat bed screen, the disc screen, and the trommel screen. Of the three, the trommel has proved to be quite effective and efficient, and hence it is the one most commonly used.

*Trommel.* The trommel is a downwardly inclined, rotary cylindrical screen. Its screening surface is either a wire mesh or a perforated plate. A diagram of a trommel screen is presented in Figure 4.12. The trommel can be used to process

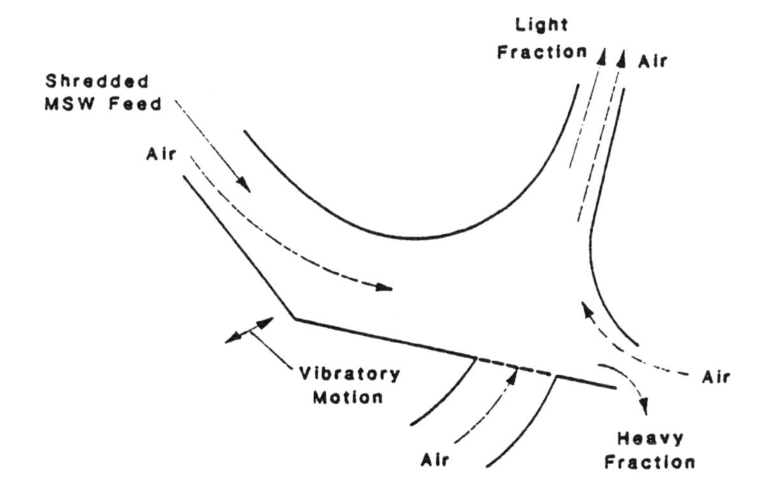


**Figure 4.9.** Vibrating inclined air classifier.

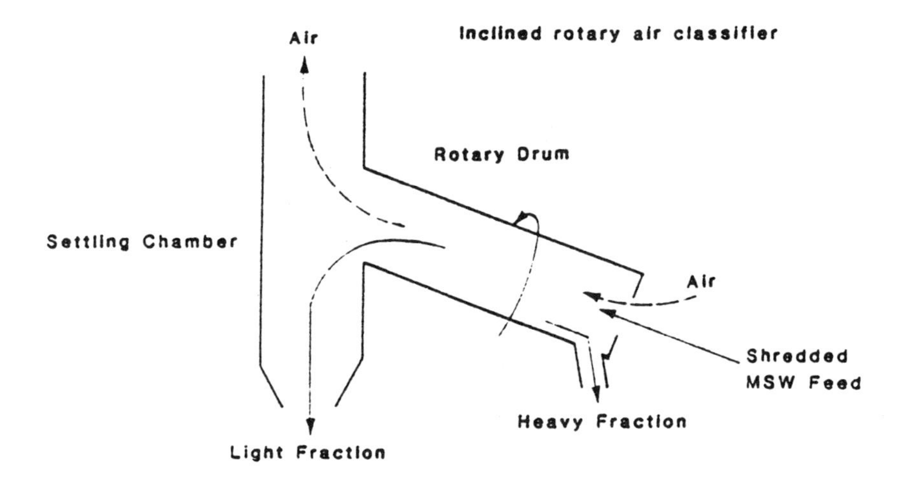


**Figure 4.10.** Inclined air classifier.

raw refuse prior to size reduction (''pretrommeling''), as well as to process shredded refuse (''post-trommeling''). With either option, the characteristic tumbling action imparted by the rotating screen results in efficient separation.[8,9]

The tumbling action efficiently separates adjoining items, or pieces of material clinging to each other, or an item from its contents. The tumbling action is essential in the screening of MSW, because of the need for a high degree of screening efficiency coupled with a minimum of screening surface.

When installed ahead of the primary shredder, the trommel can be designed to perform three important functions: (1) the removal of most of the abrasive

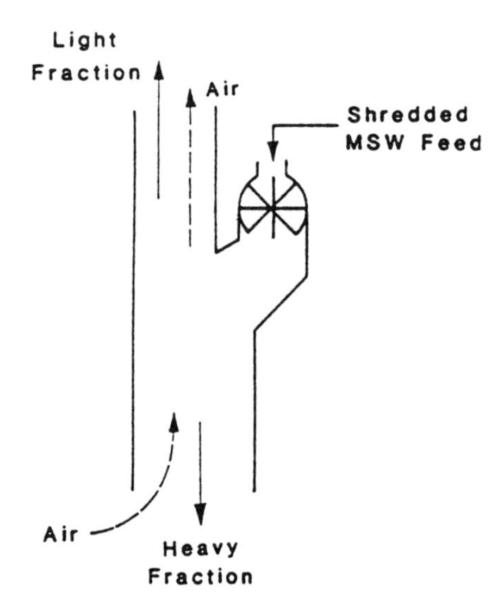


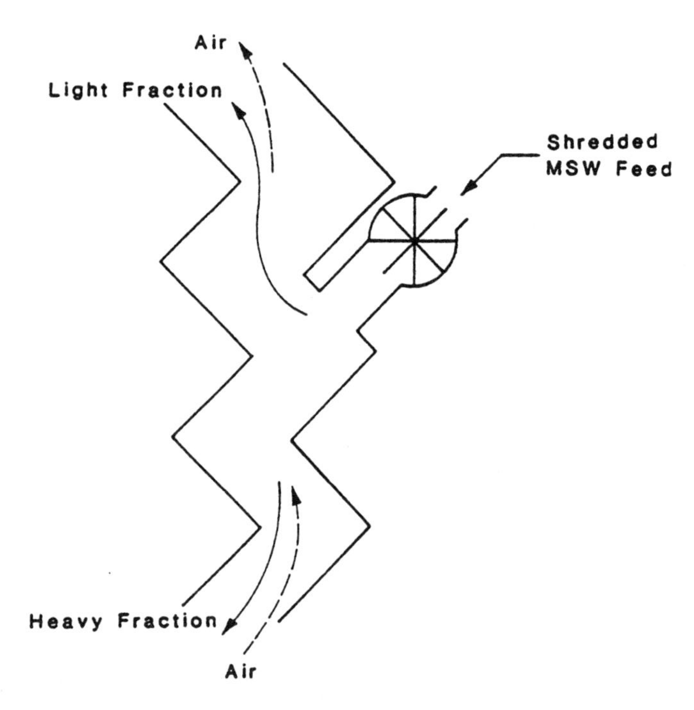


**Figure 4.11.** Types of vertical air classifier.

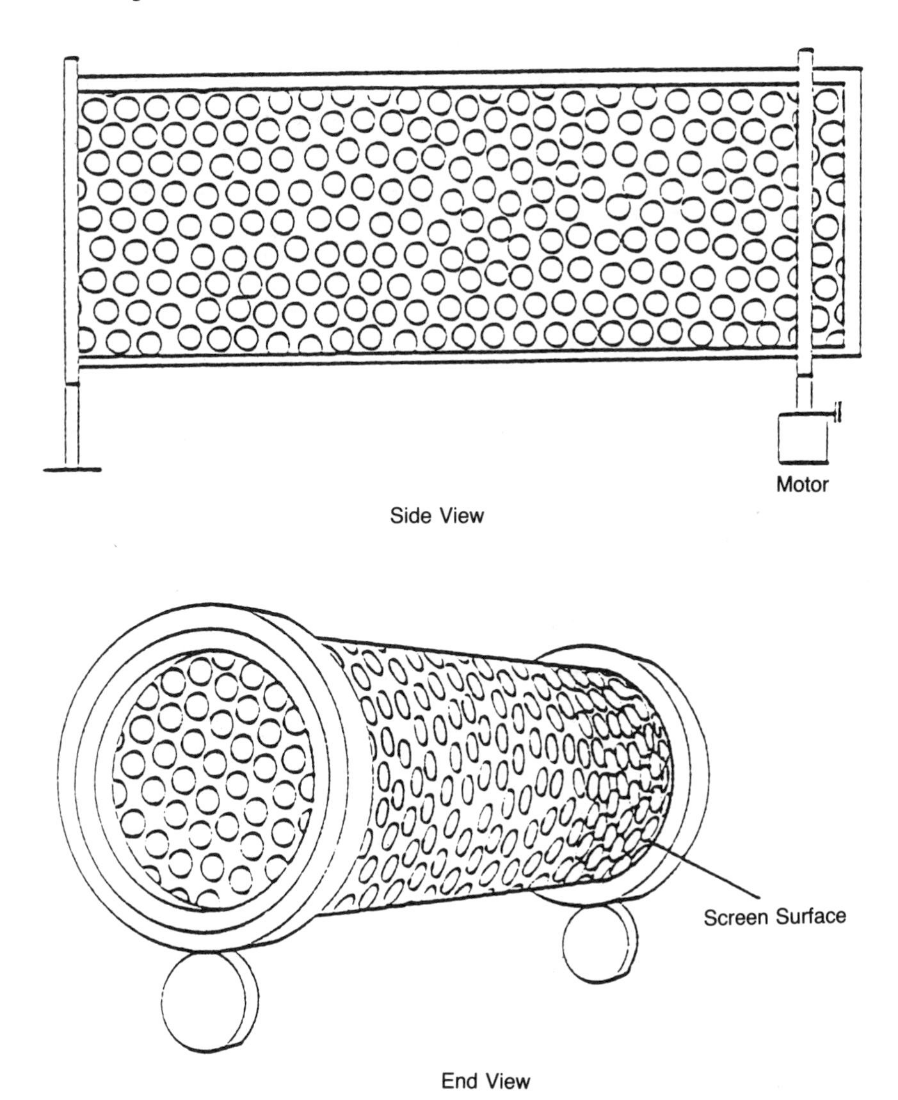


**Figure 4.12.** Trommel screen.

inorganic materials, such as dirt and stones; (2) the tearing and opening of bundles of paper and bags of waste; and (3) a coarse separation of metal, glass, and plastic containers from corrugated, ledger, and newspaper. Because of these three functions, wear and shock on shredder components are minimized, and a shredder of smaller capacity can be used downstream. If installed downstream of the primary shredder, the trommel can be designed to remove pulverized glass, dirt, heavy food particles, and other organic materials, an important function for certain recovery applications. Because the shredded particles of MSW

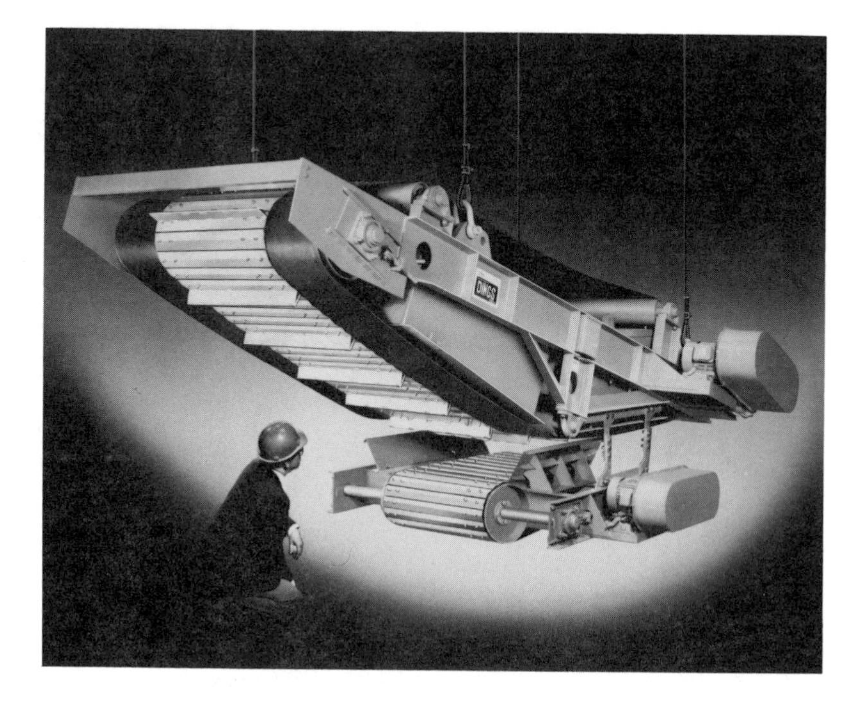


**Figure 4.13.** Magnetic belt. (Courtesy of Dings.)

generally have lesser weights than those of unshredded (i.e., raw) MSW, and because the material input is less, a smaller and lighter trommel can be used than if raw MSW is screened.

### *Magnetic Separation*

Magnetic separation is a relatively simple process used to separate ferrous metal from MSW.[1,2,9]

Magnets used for accomplishing the separation may be either the permanent or the electromagnetic type. They come in one or more of three configurations, namely, drum, magnetic head pulley, and magnetic belt. They may be assembled and suspended in line, crossbelt, or mounted as head pulleys in the material transfer conveyor. The magnetic head pulley-conveyor consists of a magnetic pulley that serves as the head pulley of a conveyor. In its operation, the material to be sorted passes over the magnetic pulley, and the magnetic particles are pulled part way around the rotating pulley while the nonmagnetic particles follow an unrestrained ballistic path in a separate direction. In the drum magnet, the electromagnetic assembly usually is mounted inside the rotating drum where the assembly remains stationary. The drum magnetic assembly can be installed in either overfeed or underfeed applications. In its simplest form, the magnetic belt consists of single magnets mounted between two pulleys that support the conveyor belt mechanism. An example of a magnetic belt is shown in Figure 4.13.

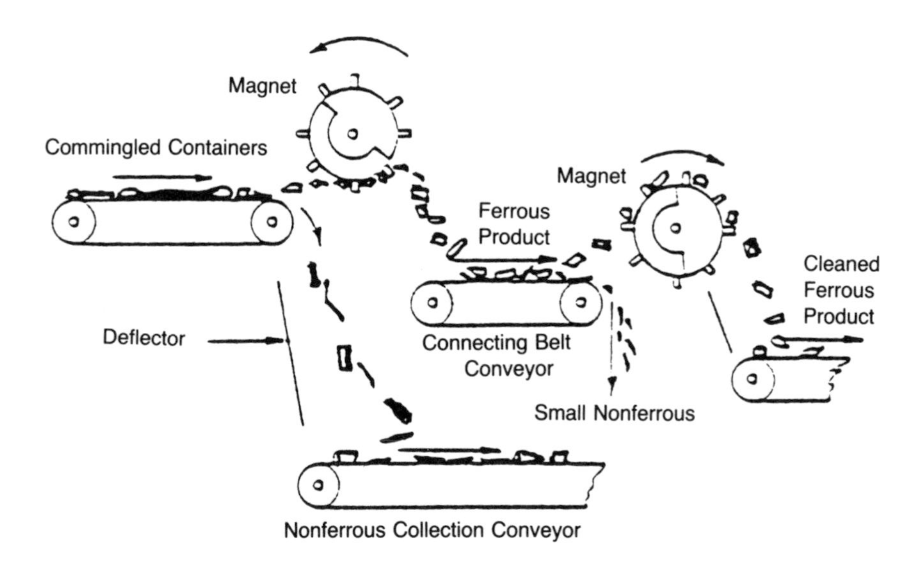


**Figure 4.14.** Multiple magnetic drum.

The concentration of magnetic metals in MSW depends upon the extent of preprocessing and recycling. In the U.S., the concentration typically can vary from about 2 to 8% by weight. In terms of weight of magnetic metal recovered per unit weight of magnetic metal in the infeed stream, the efficiency of single-stage magnetic metal recovery from shredded refuse typically is about 80%. The efficiency of ferrous metals recovery from the heavy fraction separated through air classification generally is on the order of 85 to 90%. Higher removal efficiencies can be achieved through the use of multiple magnetic separation processes, as shown in Figure 4.14.

The quality of ferrous scrap recovered from MSW by a magnetic separator placed directly downstream of primary size reduction equipment generally is inferior to that of scrap removed by a separator placed further downstream. The reason for the improvement in quality is that paper, plastic, rags, and other contaminants that otherwise might cling to or be entrapped by the ferrous scrap or be carried over with the metal, will have been removed by screening and air classification prior to the exposure of the MSW to the magnet.

### *Aluminum and Glass Separation*

Several technologically complex processes have been utilized or proposed for aluminum and glass separation. For aluminum, heavy media and eddy current separation are possibilities. Of these two processes, eddy current separation is at present the more commercially feasible. Separation is brought about by the ejection of aluminum particles from a moving waste stream due to the force exerted on the aluminum as it passes through an electromagnetic flux. Froth flotation and optical electronic sorting have both been used for glass removal. Efficient aluminum and glass recovery processes are costly and complex.[1]

## DRYING AND DENSIFICATION

Thus far, drying has been utilized at only a few waste-to-energy facilities. The objective of drying is to provide a higher quality waste-derived fuel. Because of the cost of the process and the limited success attained, drying has not generally been included in recent material processing systems. Densification — baling in particular — has been effectively used to reduce landfill requirements, and to cut transportation and disposal fees. (Tipping fees in some cases are charged by the cubic yard and not by the ton.) Because of the relatively limited processing capacity and the need to process the feedstock to an exceedingly fine particle size, densification by way of briquetting, pelletizing, or cube formation apparently is impractical for all but a few operations. Densification is used primarily for the production of a solid fuel from the light (i.e., combustible) fraction of MSW or from different paper fractions.

## REFERENCES

1. Diaz, L.F., G.M. Savage, and C.G. Golueke, *Resource Recovery from Municipal Solid Wastes: Vol. I, Primary Processing* (Boca Raton, FL: CRC Press, Inc., 1982).
2. Savage, G.M., J.C. Glaub, and L.F. Diaz, *Unit Operations Models for Solid Waste Processing* (Park Ridge, NJ: Noyes Publications, 1986).
3. Savage, G.M., et al. ''Engineering Design Manual for Solid Waste Size Reduction Equipment,'' prepared by CalRecovery Inc. for the U.S. Environmental Protection Agency, Report No. EPA-600/S8-82028 (1982).
4. Hasselriis, F., *Refuse-Derived Fuel Processing* (Stoneham, MA: Butterworth Publishers, 1984).
5. Vesilind, P.A. and A.E. Rimer, *Unit Operations in Resource Recovery Engineering* (Engelwood Cliffs, NJ: Prentice-Hall, Inc., 1981).
6. Bendersky, D., G.M. Savage, et al. *Resource Recovery Processing Equipment* (Noyes Data Corporation, 1982).
7. Robinson, W.D., ''Shredding Systems for Mixed Municipal and Industrial Waste,'' in Proc. of the American Soc. of Mechanical Engineers Conf. (1976).
8. ''Trommel Screen Research & Development for Applications in Resource Recovery,'' prepared by CalRecovery, Inc., for the U.S. Department of Energy, October (1981).
9. Robinson, W.D., Ed. *The Solid Waste Handbook* (New York: John Wiley & Sons, 1986).

# CHAPTER 5

# Recycling — MRFs

## INTRODUCTION

Recycling may play several roles in a solid waste management plan. Recycling may be the only major consideration, one among other major considerations, a minor consideration, or not a consideration in the design of a municipal solid waste management plan. At present, practically every waste management plan includes at least a minimum provision for recycling. In fact, it is a major consideration in most management plans.

Regardless of its priority and assuming the completion of all necessary preliminaries (e.g., public motivation and education, regulations), recycling begins at the point of waste generation, advances in waste collection, and eventually reaches completion at a resource recovery facility. In this book, the initiation of recycling and its continuation through storage and collection were described in special sections of Chapter 2. This chapter deals with the completion of the recycling activity at a recycling facility. The facility may be a part of the disposal facility or it may be an independent entity. In its independent form, a recycling facility has acquired the designation, "Materials Recovery Facility," which becomes "MRF" in waste management jargon.

## DESIGN OF MRFs

### Requisite Features

The design of a successful MRF incorporates certain features, among which are the following: (1) reliance is had upon proven concepts; (2) an adaptation is made not only to the characteristics of the waste from which the desired resources are to be recovered, but also to the characteristics of the resources; (3) the quality of the recovered resource is either preserved or improved; (4) processing flexibility to accommodate potential future changes in market conditions; and (5) the diversion of wastes from landfills is magnified because the feasibly largest amount of resources is recovered.[1,2]

Features pertaining to operation include provisions for (1) receiving MSW and accommodating the various types of vehicles that deliver the waste to the facility as well as at the frequency of the deliveries; and (2) relying upon manual labor when current automation technology is lacking, unproven, or only marginally effective.

Additional operation-oriented features are establishment of throughput capacity, required availability, and desired redundancy for the system.

Throughput, availability, and redundancy are critical factors in the design of any process. Unfortunately, however, many waste processing facilities are built without the benefit of using these factors. A good understanding of the quantity and composition of the feedstock to the facility allows for determining the size, type of equipment, hours of operation, quality of throughputs, and other items. Availability is the estimated amount of time that a particular piece of equipment (or system) is "available" to perform the task for which it is intended. Availability is expressed in terms of percentage. For example, a shredder is placed in a process scheduled to operate 16 hr/day. However, it is known that the shredder is expected to be out of service 4 hr (for maintenance, repairs, pluggage, etc.). Then the availability of the shredder is calculated as follows:

$$\text{availability} = \frac{16\text{-}4}{16} = 75\%$$

The importance of availability cannot be overemphasized, particularly in waste processing facilities which are both maintenance intensive and cannot be easily shut down without upsetting the solid waste management system.

Redundancy, on the other hand, is part of a "safety factor." A certain amount of redundancy must be built into a design to allow for stoppages in a particular piece of equipment or processing line. The stoppages may be scheduled or unscheduled. Although redundancy is required to maintain continuity in a particular process, redundancy has the deleterious effect of increasing capital costs. Consequently, redundancy often is ignored or minimized in order to maintain costs to a minimum.

Availability and redundancy are two factors that have played key roles in the closure of facilities. These factors are particularly important in waste processing due to the lack of reliable information on the performance of equipment and the systems used.

### Basic Consideration of Overall Design

A fundamental consideration in the overall design of a materials recovery facility is whether the input MSW is mixed or is source separated. In as much as in this division, the term "mixed MSW" is interpreted literally, and accordingly the wastes are not separated prior to collection. Obviously, the mixture contains several components. Conversely, source-separated wastes refer to wastes that have been separated into individual components at the site of generation and are collected as such, i.e., kept separated throughout collection and transport. In practice, the term has been broadened to include components in commingled form, i.e., in separate groups, each of which consists of two or more types of components. An example of a group is one consisting of metal and glass containers.

The advantages attending source separation were discussed in some detail in Chapter 2. Here it suffices to point out that the essence of the combined advantages is the avoidance of the disadvantages associated with processing mixed MSW. Thus, source-separated recyclable materials are less contaminated with food wastes and other objectionable contaminants than those present in mixed MSW. It follows that the lower degree of contamination significantly raises the percentage recovery of materials in the form of products and raises the quality of the products from source-separated wastes beyond that from mixed wastes.

A variety of material categories serves as feedstocks for MRFs. Individual categories (e.g., tin cans or glass containers) may be delivered singularly or in a myriad of commingled forms (i.e., specified mixtures of a few individual categories such as glass and metal containers). The design of the physical layout of the MRF and selection of equipment is primarily a function of (1) the quantities and composition of each of the feedstock streams that will enter the facility, and (2) the market specifications of the recovered products. Other design considerations include the potential need for and the benefits of processing flexibility. Flexibility includes the provision of producing more than one marketable form for a particular material type, e.g., baled and granulated forms of PET. An illustration of applicable design considerations and processing alternatives for a variety of source-separated MRF feedstocks, whether delivered in individually segregated or various commingled forms, is provided in Table 5.1.

A wide range of equipment is applicable to processing of materials, as illustrated in Table 5.2. The variety of equipment mirrors the variety of the forms of the feedstocks, of feedstock composition, and of market specifications that may apply to a particular MRF project. Equipment is also required for environmental control and for processing control and documentation (e.g., weigh scales).

The processing of materials and the recovery and preparation of end products to market specifications is a complex undertaking if high efficiency and production rates are the system design objectives. One of the reasons is the sheer number of factors that must be considered and optimized. A partial listing of important factors and their implication to MRF system design is given in Table 5.3.

Materials processing facilities typically require substantial manual sorting as primary and secondary separation operations. Manual sorting is obviously a labor and time intensive activity. In fact, a substantial portion of the operating costs of a MRF can be that associated with sorting labor. Sorting rates and efficiency are influenced by a number of factors, including the type and form of material to be segregated and degree of contamination and of commingling.

**Table 5.1 Typical Design Considerations and Processing Alternatives for MRFs that Process Source Separated Feedstocks**

| Collection Category | Basic Feedstock | Tipping Floor | Sorting Conveyor (or Room) | Interim Storage | Preparation for Shipping | Finished Product Storage |
|---|---|---|---|---|---|---|
| Newspaper | Newspapers, kraft bags, *retrogravture,* some coated grades | Handpick contaminants | Handpick contaminants | Accumulated in bins or bunkers before being selectively conveyed to baler | Baler | In stacks or bales on processing floor or stacked in transport vehicles |

| Collection Category | Basic Feedstock | Tipping Floor | Infeed Conveyor | Screen | Dynamic/Pneumatic Separator |
|---|---|---|---|---|---|
| Commingled containers | Tin, bimetal, aluminum cans, plastic and glass containers, contaminants | Handpick contaminants | Handpick contaminants; magnetic separator for ferrous | Broken glass recovered as undersized mixed-color fraction | Separate aluminum and plastic from glass |

| Collection Category | Sort Method | Bale | Biscuit | Shred | Air Classify | Store |
|---|---|---|---|---|---|---|
| Ferrous (bimetal) | Magnetic and/or manual separation of tin cans and bimetal (if required) | With baler | With can densifier | With can shredder | | Convey shredded cans to outside transport vehicle, or bales or biscuits in stacks on processing floor, outdoors, or in a transport vehicle |
| Ferrous (tin cans) | Magnetic and/or manual separation of tin cans and bimetal (if required) | With baler | With can densifier | With can shredder | To remove labels | Convey shredded cans to outside transport vehicle, or bales or biscuits in stacks on processing floor, outdoors, or in a transport vehicle |

| Collection Category | Sort Method | Flatten | Store | Bale | Biscuit | Store |
|---|---|---|---|---|---|---|
| Aluminum | Eddy current apparatus separates aluminum from nonmetals | With can flattener | Pneumatically convey to outside transport vehicle | With baler | Compress in a densifier | On the processing floor or outdoors or in a transport vehicle |

| Collection Category | Sort Method | Interim Storage | Perforate | Bale | Store |
|---|---|---|---|---|---|
| Plastic (PET) | Pneumatic and/or manual sort of PET | In overhead hoppers | Drop from overhead hopper or pneumatically convey to perforator | Mechanically or pneumatically from perforator to baler | On processing floor or outdoors in transport vehicles |

| Collection Category | Sort Method | Interim Storage | Granulate | Bale | Store |
|---|---|---|---|---|---|
| Plastic (HDPE) | Manual sort of HDPE | In overhead hoppers | Drop from overhead or pneumatically convey to granulator | Mechanically or pneumatically convey to baler | Granulated in gaylords on processing floor before loading into transport vehicle, baled in stacks on processing floor or outdoors in transport vehicles |

| Collection Category | Sort Method | Crush | Upgrade | Store |
|---|---|---|---|---|
| Glass | Optical automatic sort or hand sort by color | With glass crusher | Remove paper labels, metal lids, and other contaminants by screen and/or air classifier | In bunkers for loading by front-end loader, or in overhead bins for selectively conveying to transport vehicles |

| Collection Category | Sort Method | Bale | Store |
|---|---|---|---|
| Plastic (HDPE and PET) | Manual sort of each type of resin | Mechanically or pneumatically convey to baler | In bunkers for loading by front-end loader, or in overhead bins for selectively conveying to transport vehicles |

**Table 5.2 Fixed Equipment Applicable for Materials Recovery Facilities**

| |
|---|
| Material handling equipment |
| Belt conveyor |
| Screw conveyor |
| Apron conveyor |
| Bucket elevator |
| Drag conveyor |
| Pneumatic conveyor |
| Vibrating conveyor |
| Debagger |
| Separating equipment |
| Magnetic separator |
| Eddy current device (aluminum separator) |
| Disc screen |
| Trommel screen |
| Vibrating flat-bed screen |
| Traveling chain curtain |
| Air classifier |
| Size reduction equipment |
| Can shredder |
| Can densifier/biscuiter |
| Can flattener |
| Glass crusher |
| Plastics granulator |
| Plastics perforator |
| Baler |
| Environmental control equipment |
| Dust collection system |
| Noise suppression devices |
| Odor control system |
| Heating, ventilating, and air conditioning (HVAC) |
| Other equipment |
| Fixed storage bin |
| Live-bottom storage bin |
| Floor scale for pallet or bin loads |
| Truck scale |
| Belt scale |

Ranges of sorting rates and of recovery efficiencies can be established that cover the usual range of operating conditions at MRFs. Ranges for selected material categories are presented in Table 5.4.

In addition to sorting labor, the operational staff of MRFs usually includes operators of fixed and rolling equipment and maintenance personnel. Sorting labor, as mentioned previously, can be a substantial portion of the operating staff, sometimes 50 to 75%. MRF facilities also require an office staff that may include one or more of the following: plant manager, weigh master, bookkeeper, clerk, custodian, etc. Typical staffing levels for MRFs of several selected processing capacities are illustrated in Table 5.5.

One aspect of MRF design deserves special attention. That aspect is space allocation for storage and processing of materials. Storage areas include those allocated for tipping floor storage of delivered materials and for storage of

**Table 5.3 Some Factors Affecting Process Design and Efficiency**

| Factor | Design Implication |
|---|---|
| Market specifications | Loosely constrained specifications generally result in percentage yields that are higher than those for tightly constrained specifications for the end products |
| Contamination of incoming materials | The greater the degree of feedstock contamination, the lower the percentage yield of product for a given set of end product specifications |
| Glass breakage | Broken glass containers are more difficult to sort than unbroken containers |
| Relative quantities per sorter | Over a given period of time the greater the number of units and/or components a sorter must separate from a mixture the greater the rate of error and conversely, the lower the recovery rate |
| Equipment design | The proper design of conveyors and separation equipment for the types and quantities of materials handled directly affects recovery rates; for example, an excessive bed depth of commingled containers on a conveyor can substantially limit the manual or automatic recovery efficiency of any given component |
| Human factors | Providing a clean, well-lighted, well-ventilated environment in which to work with particular attention to worker training, safety, health, and comfort is conducive to high recovery rates |

**Table 5.4 Manual Sorting Rates and Efficiencies**

| Material | Lb/Hr/Sorter | Recovery Efficiency (%) |
|---|---|---|
| Newspaper[a] | 1,500–10,000 | 60–95 |
| Corrugated[a] | 1,500–10,000 | 60–95 |
| Glass containers[b] (mixed color) | 900–1,800 | 70–95 |
| Glass containers[b] (by color) | 450–900 | 80–95 |
| Plastic containers[b] (PET, HDPE) | 300–600 | 80–95 |
| Aluminum cans[b] | 100–120 | 80–95 |

[a] From a processing stream of predominantly paper of one or more grades.

[b] From a processing stream of predominantly metal, glass, and plastics, e.g., commingled metal and glass containers.

**Table 5.5 Typical Staffing Requirements for Material Recovery Facilities**

| | Tons per Week | | |
|---|---|---|---|
| Personnel | 500 | 1000 | 2000 |
| Office | 2–3 | 3–4 | 5–6 |
| Operational | | | |
| Foreman/machine operator | 1–2 | 2–3 | 3–4 |
| Sorters | 8–16 | 10–18 | 15–26 |
| Forklift/FEL operators | 2–3 | 3–4 | 5–6 |
| Maintenance | 1 | 2 | 4 |
| TOTAL | 14–25 | 20–31 | 32–46 |

Table 5.6 MRF Floor Area Guidelines (Square Feet)

| Area Use | Capacity (TPD) 10 | 100 | 500 |
|---|---|---|---|
| Tipping floor | | | |
| 2-day capacity | 3,000 | 7,500 | 30,000 |
| Processing | 6,000 | 20,000 | 50,000 |
| Storage | | | |
| 7-day capacity | 1,000 | 8,750 | 35,000 |
| 14-day capacity | 1,750 | 17,500 | 70,000 |

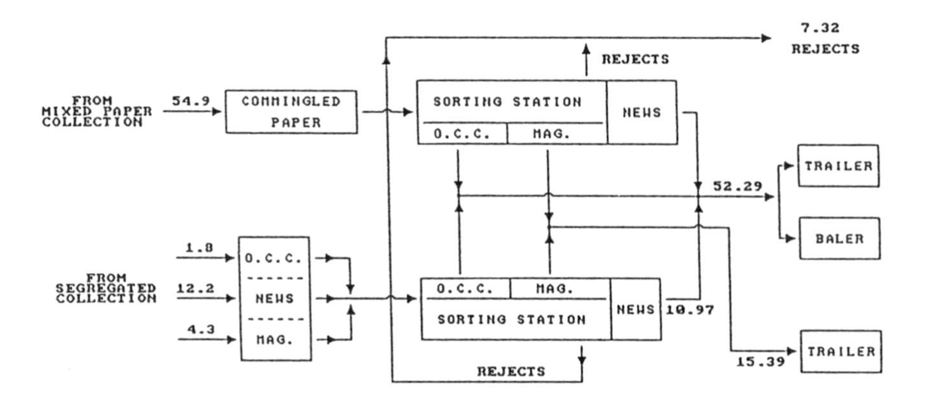


Figure 5.1. Paper processing line/design capacity = 75 TPD; 75% commingled collection, 25% segregated collection.

recovered end products. The usual tendency is to underestimate space requirements, with the potential results being loss of processing flexibility and unprotected storage of materials (e.g., outdoors) because of lack of adequate storage space indoors, or both. Some general guidelines for space allocation are presented in Table 5.6. The guidelines are not meant to be a substitute for a detailed engineering analysis that considers actual throughputs and other project-specific criteria.

Two designs have been selected as examples of the influence of source-separated and of mixed MSW input on MRF design. Both designs have been reported in detail in Reference 2.

## MRF for Processing Source-Separated MSW

The first example is a 125-TPD facility designed for processing source-separated MSW. For the example, it is assumed that 25% of the recyclables arrive at the facility in presegregated, single form (e.g., tin cans) and that the remaining 75% is commingled. A process flow diagram for the paper processing line in the facility is shown in Figure 5.1. Figure 5.2 is a process flow diagram for a container processing line in the same facility. The flow diagrams also serve as

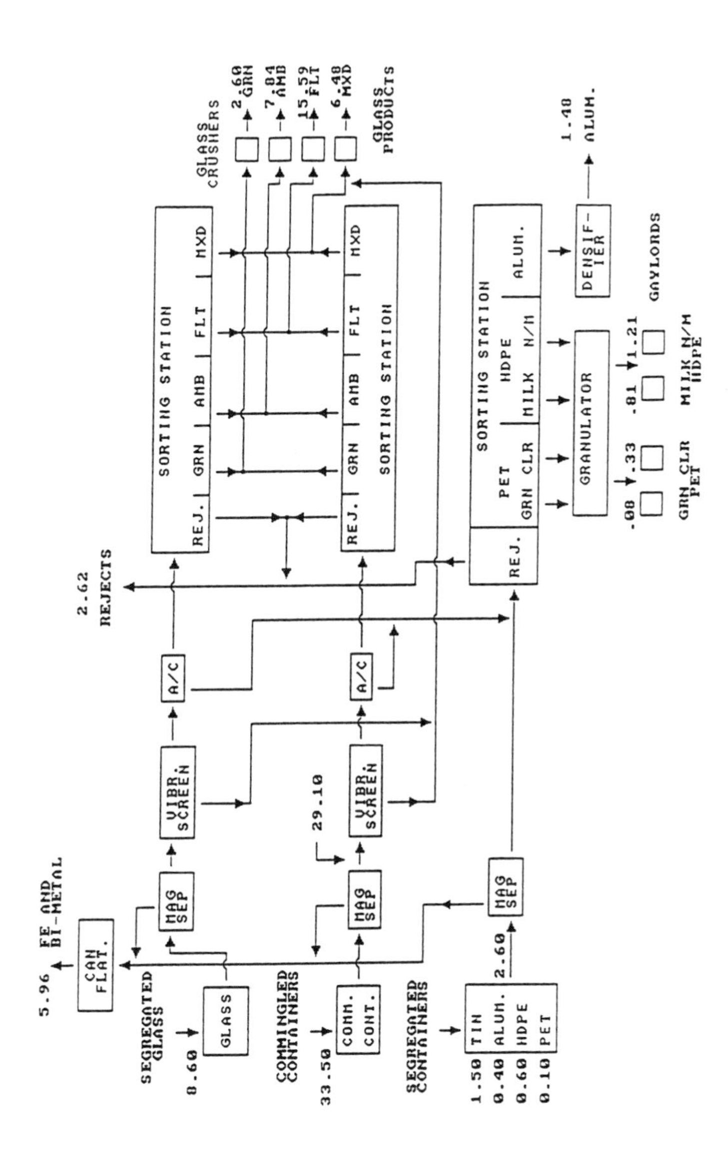


**Figure 5.2.** Container processing line/design capacity = 45 TPD; 75% commingled collection, 25% segregated collection.

mass balances, showing the tonnages of the various recyclables introduced into the system and discharged from it. Both flow diagrams show provision for redundancy in receiving, sorting, and processing.

Breakage and contamination generally is about 2 to 12% of the material introduced into such a system. Glass breakage during collection and during processing at the facility results in the loss of small particles of glass as residue. This residue is lost if a market for mixed colored cullet cannot be found.

Contaminants must be removed to the extent indicated by market specifications. With respect to paper and paper products recycling and processing in MRFs, common contaminants include (1) corrugated and magazines that advertently or inadvertently had been intermingled with the residential newspaper prior to or during collection; and (2) low-grade paper (e.g., envelopes with ''windows'') that had been intermingled with commercial high-grade paper before or during collection.

A plan view of a facility designed to accommodate the processing lines diagrammed in Figures 5.1 and 5.2 is presented in Figure 5.3. The tipping floor and product storage areas are sufficiently large to supply at least one day of storage for all materials. Moreover, the facility would provide extensive redundancy and flexibility in the two lines. As shown in Figure 5.3, the tipping floor for the paper fraction has two receiving pits, each of which serves a processing line that could handle either the entire anticipated input of mixed paper waste or the entire anticipated input of segregated paper waste. Thus, each pit serves as a backup for the other.

The container tipping floor has three receiving pits. Two of the pits and their associated processing lines are completely redundant in that each pit and associated line can process either the entire anticipated input of mixed container waste or the entire anticipated input of segregated container waste. The third line is intended solely for handling segregated plastic and aluminum containers.

### *Evaluation*

Although the extensive redundancy provided by the design would ensure a minimum risk of downtime due to equipment failure, it also would render the design expensive to put in practice. Consequently, substantial savings would be realized by eliminating redundant processing capability. The elimination could be compensated by operating the facility on at least a two-shift basis. Nevertheless, curtailment of redundancy must be accompanied by the establishment of standby plans for minimizing the effects of outages resulting from the breakdowns of machinery that are inevitable in all plants.

## MRF for Mixed MSW

MRFs can recover 80% or more of the marketable grades of metals, glass, plastics, and paper from an input consisting of commingled and segregated components. On the other hand, MRFs processing mixed MSW usually recover

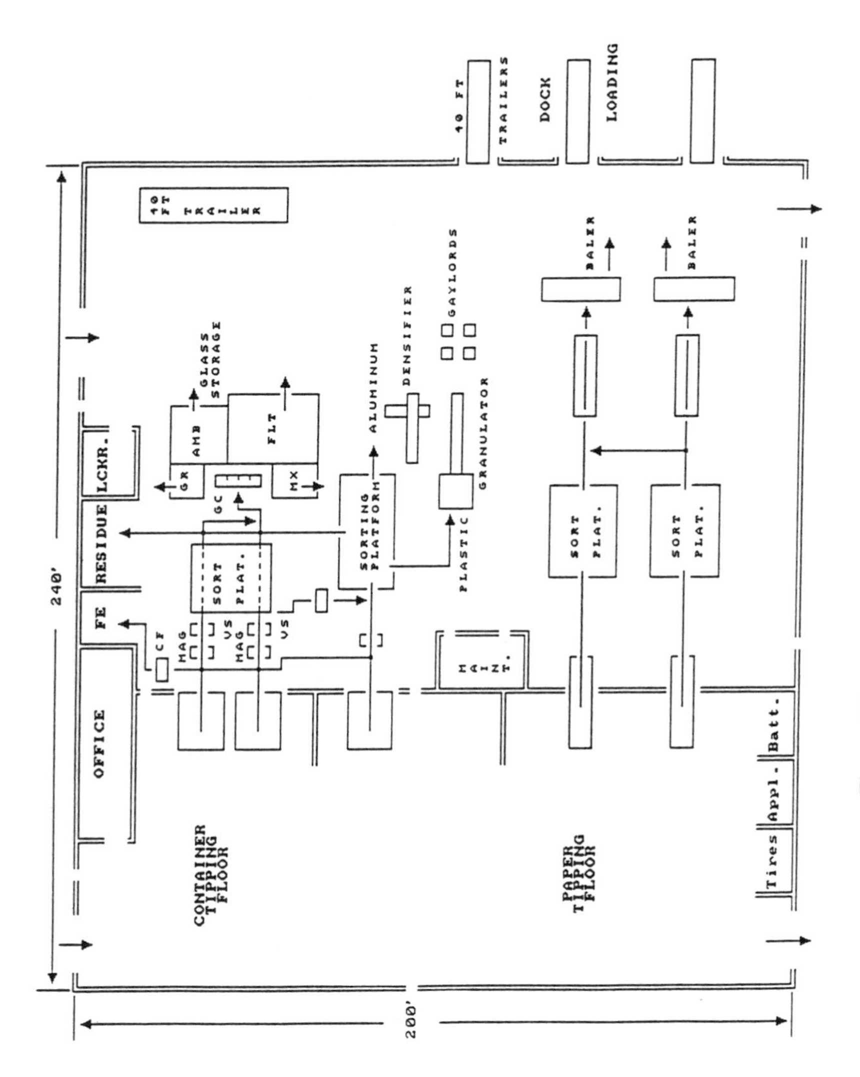


**Figure 5.3.** Plan view of materials recovery facility.

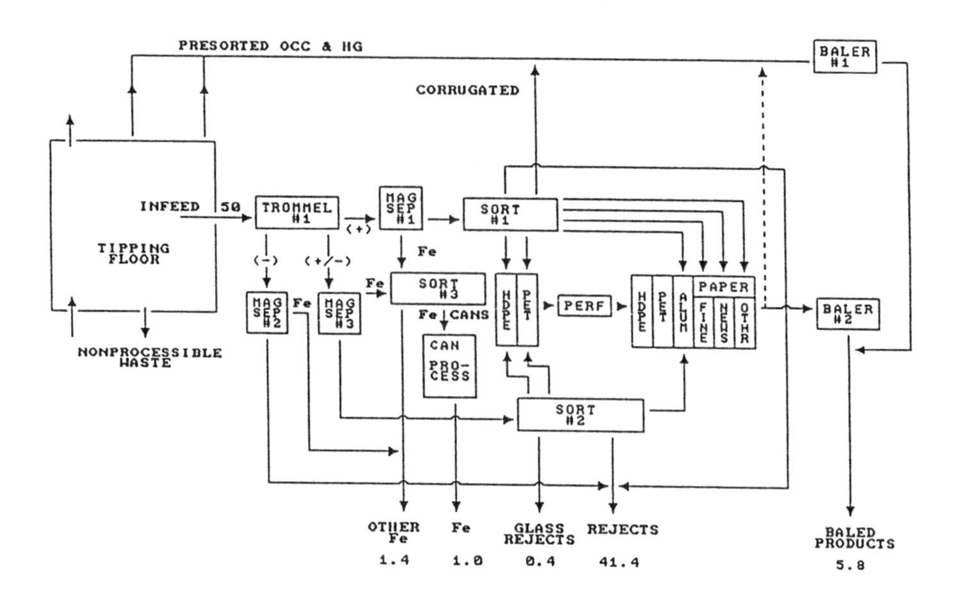


**Figure 5.4.** Flow diagram for a mixed MSW MRF (TPH).

only about 10 to 20% of the wastes in the form of marketable materials. However, by integrating into the design of the latter facilities, additional processing operations to recover refuse-derived fuel (RDF) or a compostable feedstock, it is possible to increase their resource recovery capability to 75 to 85%, provided that there are markets for the two materials.

Figure 5.4 is the flow diagram of a 50-TPH materials recovery facility design. The design is configured primarily to process and recover recyclable materials (including ferrous, high density polyethylene (HDPE), polyethylene teraphthalate (PET), aluminum, and several grades of paper from mixed MSW. So as to optimize the recovery of marketable secondary materials, the processing system relies upon both mechanical and manual separation. The design permits the recovery of approximately 15% of the input mixed waste in the form of marketable grades of recyclables. It is assumed that wastes are delivered to the facility by means of conventional refuse collection vehicles and/or transfer trailers.

The design of the facility, of which Figure 5.4 is the flow diagram, calls for wheel loaders and a picking crane to remove large, heavy objects and other nonprocessible items from the incoming MSW before it is introduced into the processing equipment. Nonprocessible items typically include plastic strapping and film sheeting, long lengths of pipe, large automotive parts, carpets, etc. Additionally, provision is made to use wheel loaders in the facility to segregate corrugated and other marketable waste paper grades by wheel loader from incoming loads consisting predominantly of paper materials. When a sufficient amount of corrugated or other paper grades has been removed and has been accumulated on the tipping floor by wheel loader, the material is transferred directly to a baler, thereby bypassing the mixed waste processing equipment.

As provided by the design, mixed MSW is introduced into a two-stage primary trommel. Ferrous metal is extracted from the first stage trommel undersize by passing it by a magnetic separator. Residue from this processing is routed to the output residue stream. The second-stage undersize from the primary trommel is passed through a magnetic separator. Ferrous metal thus removed is conveyed to a sorting station (Sort #3 in Figure 5.4). At the sorting station, ferrous metal that has been magnetically extracted from the trommel oversize material is added to the ferrous metal from the second-stage trommel undersize. Ferrous cans are sorted from other ferrous materials and are discharged into a can processing subsystem. The resulting product would be minimally contaminated.

After having been subjected to magnetic separation, the primary trommel oversize is conveyed to another sorting station (Sort #1 in Figure 5.4). At this station, HDPE, PET, aluminum, cardboard, and various paper grades are separated manually. The separated materials are baled by means of one of two balers — the second baler serves as a processing redundancy.

HDPE, PET containers, aluminum, and some high-grade paper are manually sorted at the third sorting station. This station receives undersize from the second stage of the primary trommel after ferrous removal. The remaining waste is added to the waste from the sorting station at which the trommel oversize stream is processed.

Manual sorting is relied upon to a considerable extent for segregating plastics and aluminum because it is an efficient means for segregating and recovering the various plastic polymers and aluminum beverage containers, and because it provides an opportunity for employment development. Moreover, the use of mechanical and electromechanical separation systems for plastic polymers and aluminum materials in waste processing is as yet in the developmental stage.

Much of the process residues that account for about 85% of the incoming solid waste is either combustible or is biodegradable organic material. Unless these residues are processed for use in energy recovery or are converted into a feedstock for composting, they must be landfilled. However, integration of the organic residues with refuse-derived fuel recovery could reduce the size of the residue stream to about 15 to 25% of the input MSW.

## Summary

### *Types of Materials (Feedstocks)*

MRFs generally are designed to handle the following types of feedstock materials:

1. Presorted (i.e., source separated) individual recyclable components
2. Presorted commingled waste components
3. Unsorted mixed special waste having a dominant recyclable material

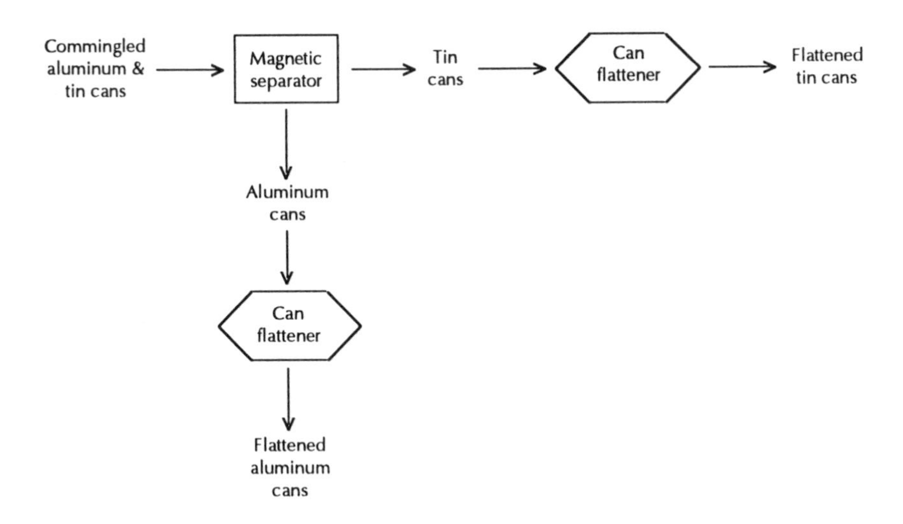


**Figure 5.5.** Commingled cans processing.

The following are examples of each of the three feedstocks: (1) individual components such as paper, aluminum cans, glass bottles, and yard waste; (2) commingled components such as paper and plastic; glass containers and metal cans; and steel and aluminum cans; and (3) mixed special waste such as commercial mixed waste composed predominantly of paper and contaminated with plastic, ferrous, aluminum, and glass.

### *Manual and Mechanical Processing*

Sorting and separation in MRFs usually involves both manual and mechanical processing. Generally, manual sorting is used to separate the corrugated, newsprint, and plastic components in the oversize fraction of incoming wastes. If their quantities warrant doing it, individual components may be baled.

Three representative uses of mechanical processing in MRFs are processing commingled beverage cans, processing commercial mixed wastes, and processing yard waste.

*Processing Commingled Beverage Cans.* Mechanical processing is begun by subjecting the cans to magnetic separation to sort ferrous material from nonferrous metal, particularly the aluminum cans. Separated ferrous cans are flattened in a machine designed for the purpose. Separated aluminum cans also may be flattened or baled. The flattening is done to conserve on storage space. The sequence of the steps is indicated in Figure 5.5.[1]

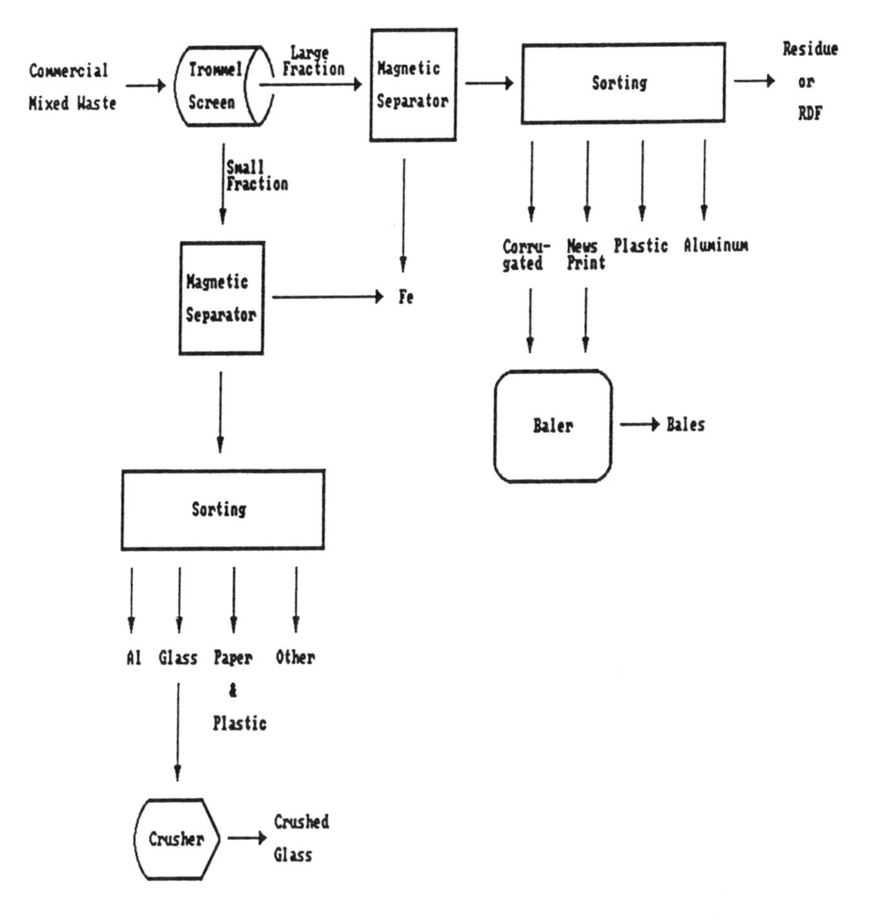


**Figure 5.6.** Commercial mixed waste processing.

*Processing Mixed Commercial Waste.* As is indicated in Figure 5.6, mechanical processing of commercial mixed waste may be initiated by introducing the waste into a screening device, such as a rotating trommel screen. If the screen is appropriately designed, virtually all of the glass is discharged as part of the "undersize." On the other hand, most of the corrugated is retained on the screen surface as "oversize." Undersize and oversize from the screening are each separately subjected to magnetic separation.

## Conclusions

Several considerations enter into the design of a materials recovery facility. One such consideration is the form of the delivered feedstock — e.g., source-separated recyclables vs mixed MSW. Another key consideration is the required extent of recycling or waste diversion. Some 10 to 20% diversion may be attained through a residential and commercial source separation program (i.e., collection plus processing) for paper grades and glass, metal, and plastic containers, whereas the inclusion of mixed waste processing and source-separated yard waste processing may be required for diversion goals of 20% or greater. Of course, in both cases, markets must be available for the recovered products.

Regardless of the degree to which the incoming waste stream is specified to be presorted at the waste generator site, it must be recognized that the waste will almost inevitably be contaminated with materials other than those specified. Historically, in facilities processing residential source-separated materials, process residues are typically 2 to 12% of the quantities processed. Manual sorting is usually provided to handle this eventuality. Similarly, each piece of mechanical processing equipment may extract material other than that which was planned. Extraneous material may become entrapped or entrained with the desired separated material and may have to be manually removed.

A more detailed presentation on the design of MRFs is given in Reference 11.

## YARD AND FOOD WASTES

Because of the potential for an MRF to receive and treat yard and food wastes, a discussion on these materials is presented here. This section should be read in conjunction with Chapters 6, 7, and 9.

### Yard Waste

"Yard waste" is the term applied to the variety of wastes of plant origin generated in the course of gardening, landscaping, and general grounds maintenance. Sources of yard waste may be public, residential, institutional (public or private), and commercial sectors. Public sources include parks, public gardens, and landscaping (initiation and maintenance) of public properties. The composition and quantities of yard debris of residential origin are influenced by several factors, chief of which are location and type of the residential area. Residential sources include single-family residences and multifamily units (apartment and condominium complexes). Residential units in rural settings generate more varied and larger amounts of debris than those in suburban areas — and far more than those in inner-city situations. Volume and types generated in institutional and commercial park settings are fairly similar. Not to be overlooked is curbside landscaping, of which trees are a major constituent.

#### *Components (Composition) of Yard Waste*

Among the principal yard waste components of concern in collection, management, and disposal (treatment and final) are (1) leaves (especially from

**Table 5.7 Approximate Concentration of Nitrogen in Yard Wastes**

| Component | Nitrogen (Percent Dry Weight) |
|---|---|
| Grass clippings | 2.15–4.5 |
| Leaves | 0.5–1.0 |
| Sawdust | 0.11 |
| Wood (pine) | 0.07 |
| Fruit wastes | 1.52 |
| Paper | 0.25 |
| Table scraps | — |

*Source:* Sills and Carrow;[3] Rosen et al.[4]

deciduous shrubs and trees); (2) discarded herbaceous plants or trimmings thereof; (3) trimmings and branches (large and small) of large shrubs, ornamentals, and trees; and (4) grass clippings. These components differ one from the other with respect to physical and chemical (nutrient) properties and to biodegradability. For example, the leaves collected in autumn are highly carbonaceous and have very little nitrogen. The structural nature and the carbon to nitrogen ratio of freshly discarded "green" herbaceous plants and trimmings are conducive to rapid decomposition, whereas those of mature (ripened) are not. For example, stalks of actively growing corn vs those of mature (ripe) corn. Another example is the high lignin content of large tree branches vs those of growing twigs. All of these characteristics affect (1) ease and type of management and disposal and thereby determine method of handling; and (2) overall physical and chemical content of the yard waste to be processed at a central facility. Approximate nitrogen contents of components that may be encountered in yard waste are listed in Table 5.7. The nitrogen content of the yard waste depends upon the relative amounts of these components present in the waste. The usual proportions of the components is such that the nitrogen content of yard waste is likely to be within the range of 1.5 to 2.0%.[4]

Ranges of concentrations of metals are presented in Table 5.8. Concentrations of metals, of microorganisms of public health significance, and of plant nutrients in composted yard waste produced at two particular sites are listed in Tables in 5.9 to 5.11, respectively.

Two reasons make it essential that a determination of the variation in rate, magnitude, and composition of yard waste with change in season be made as a prelude to the designing of a yard waste management program. The reasons are (1) the factors and conditions discussed in the preceding paragraphs vary markedly from location to location and with change in season; and (2) the factors also determine the rate, magnitude, and composition of the yard waste stream. The extent of the variations is indicated by the breadth of the ranges of the values listed in Tables 5.7 and 5.8.

Because street sweepings are likely to contain objectionable materials such as bottle caps, plastics of various types, and glass shards, they should not be mixed with yard waste destined for composting. However objectionable these materials may be, of much more serious concern are the dust and dirt in the sweepings.

**Table 5.8 Ranges of Concentrations of Metals in Plants and Yard Waste**

| Element | Cultivated Plants | Yard Waste |
|---|---|---|
| Al (%) | 0.02–0.40 | 0.06–0.31 |
| As | | |
| Ba | 15–450 | |
| Be | | |
| B | 37–540 | <0.1–1.4 |
| Cd | 0.37–2.3 | 3–14.3 |
| Cl | | |
| Cr | 0.42–6.6 | 1.2–52.5 |
| Cu | 21–230 | |
| Fe (%) | 0.06–0.27 | 0.06–0.31 |
| Pb | 7.1–87 | 1–38 |
| Mn | 96–810 | 23–1261 |
| Hg | | |
| Ni | 2.7–130 | 1.7–33.3 |
| Se | 0.04–0.17 | |
| Ag | | |
| Zn | 180–1,900 | 39–585 |

*Source:* Rosen et al.;[4] Epstein and Engel;[5] Brownio.[6]

**Table 5.9 Concentration of Soluble Metals in the Yard Debris Composts (Saturated Media in Parts per Million)**

| | Yard Debris | |
|---|---|---|
| | Site 1[a] | Site 2[b] |
| Calcium | 50 | 59 |
| Magnesium | 16 | 23 |
| Iron | 3.70 | 3.70 |
| Manganese | 0.80 | 2 |
| Zinc | 0.14 | 0.17 |
| Copper | 0.08 | 0.07 |
| Boron | 0.20 | 0.20 |
| Sulfur | 12 | 6 |
| Sodium | 21 | 31 |
| Aluminum | 4.80 | 3.30 |

*Source:* Epstein and Engel;[5] Brownio;[6] Portland Metropolitan Service District.[7]

[a] One sample.
Average of seven samples.

The dust and dirt are apt to be laden with potentially hazardous inorganic and organic substances.

It should be pointed out that in terms of content and origin, the dust in street sweepings differs only slightly from that which accumulates on the foliage of curbside plants, shrubs, and trees — particularly where the traffic is heavy. Thus, the influx of leaves during the fall may include an appreciable amount of contaminated dust.

## *Management, Treatment, and Disposal of Yard Waste*

Three principal courses of action are available, namely (1) process all yard waste at a central facility; (2) process and dispose all yard wastes at their

**Table 5.10 Concentrations of Pathogens Found in Yard Debris Compost**

| | Yard Debris | |
|---|---|---|
| **Test** | **Site 1** | **Site 2** |
| Salmonella | Neg | Neg |
| *Escherichia coli* | $>1.0 \times 10^3$ | $<1.0 \times 10^4$ |
| Fecal coliform | $2.3 \times 10^3$ | $9.3 \times 10^4$ |
| Total coliform | $1.4 \times 10^3$ | $3.0 \times 10^5$ |
| *Aspergillus fumigatus* | Neg | Neg |
| Human parasitic ova | Neg | Neg |
| Dog parasitic ova | Neg | Neg |
| *Entamoeba coli* | Neg | Neg |
| *Entamoeba histolytica* | Neg | Neg |
| *Pseudomonas* spp. | Positive | Positive |
| *Ascaris lumbriocoides* (roundworm) | Neg | Neg |
| *Taenia* spp. (tapeworm) | Neg | Neg |
| *Trichuris trichuria* (hookworm) | Neg | Neg |

*Source:* Epstein and Engel;[5] Brownio;[6] Portland Metropolitan Service District.[7]

respective sites of generation; and (3) combine courses 1 and 2, (i.e., central facility plus backyard composting). The second course would eliminate all collection of yard waste and would involve on-site (''backyard'') disposal.

Many compelling objections can be brought against the second course, i.e., no collection and no landfill disposal. The outcome is compulsory backyard processing. An obvious objection is the physical impossibility of backyard disposal in ''inner city'' situations. Another objection is the shortage of available space that is characteristic of the average city lot. Yet another overwhelming objection is the impossibility of fitting in the burden of performing the tasks associated with backyard composting into the overcrowded schedule of a household in which husband and wife are the wage earners. A final and very serious objection is that through sheer desperation on the part of the householder, backyard processing may degenerate into backyard ''landfilling'' and increased roadside littering.[8]

On the basis of the objections presented in the preceding paragraph, it becomes apparent that the best management and disposal approach is to process yard waste at a central facility and to keep backyard processing a strictly voluntary undertaking on the part of the householder — and, nevertheless, to encourage ''backyard'' composting.

### *Backyard Processing (Composting)*

The processing of choice for the backyard situation is composting, although some debris can be spaded directly into the ground prior to planting — usually in early spring or in late autumn. Principles and procedures for composting are those described in Chapter 7.

**Table 5.11 Nutrient Content and Other Parameters of Composted Yard Debris[a]**

| | | Yard Debris | |
|---|---|---|---|
| | Units | Site 1 | Site 2 |
| Total (acid digestion) | | | |
| CEC[b] | meq/100 g | 26.8 | 28.2 |
| Nitrogen | % | 0.90 | 0.63 |
| Sulfur | % | 0.26 | 0.20 |
| Phosphorus | % | 0.16 | 0.14 |
| Potassium | % | 0.72 | 0.62 |
| $NO_3$-N | % | | |
| $NO_3$-N | % | | |
| Water soluble | | | |
| Nitrogen | ppm | 2.0 | <1.0 |
| Sulfur | ppm | 12.0 | 6.0 |
| Phosphorus | ppm | 143[c] | 121[d] |
| Potassium | ppm | 3132[c] | 2604[d] |
| $NH_4$-N | ppm | 21[c] | 20[d] |
| $NO_3$-N | ppm | 6[c] | 4[d] |
| Bulk density | lb/cu yd | 594.6[c] | 726.2[d] |
| Moisture content | % | 48.5[c] | 48.9[d] |
| Organic matter | % | 67.3[c] | 64.5[d] |
| pH | | 7.1[c] | 6.7[d] |
| Specific conductance | mmho/cm[e] | 1.4[c] | 1.4[d] |
| Volatile solids | | | |
| Particle size | | | |
| 9.51 mm | % passing | 94.2[c] | 95.0[d] |
| 6.35 mm | % passing | 85.2[c] | 87.9[d] |
| 4.75 mm | % passing | 78.6[c] | 79.7[d] |
| 2.38 mm | % passing | 59.7[c] | 61.9[d] |
| 1.0 mm | % passing | 32.8[c] | 36.9[d] |
| 0.5 mm | % passing | 17.1[c] | 21.5[d] |

*Source:* Epstein and Engel;[5] Brownio;[6] Portland Metropolitan Service District.[7]

[a] Average of 10 samples.
[b] CEC = cationic exchange capacity, expressed in miliequivalents (meq) exchangeable cations per 100 g of dry soil. The CEC determines physical properties of soils.
[c] Average of eight samples.
[d] Average of five samples.
[e] mmho/cm = millimho per centimeter (a mho is a unit of conductance used to measure soluble salts).

*Discussion.* There is considerable debate as to the utility or even practicality of the prefabricated compost units presently on the market. To individuals who are well versed in the art of composting, the units seem to be too small to permit the self-insulation required for a significant accumulation and retention of heat. Apparently, the minimum volume for satisfactory heat retention is about 1 cu yd. Preferably, but not necessarily, the compost bin should be constructed of a durable material. Wood, concrete, or cement blocks are suitable.

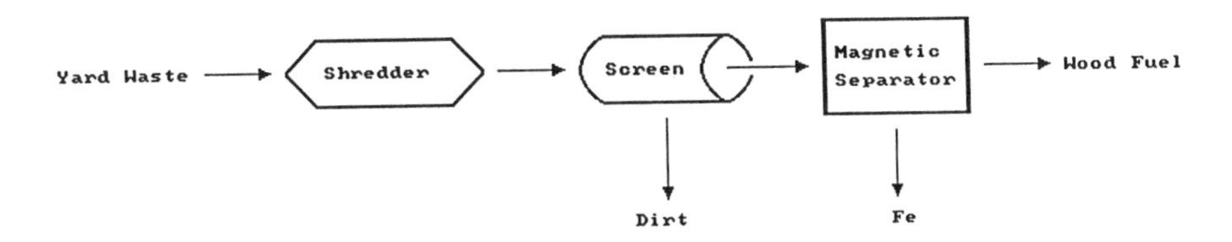


**Figure 5.7.** Yard waste processing.

Inoculums, enzymes, and other exotic additives serve no useful purpose.[9,10] However, the addition of a chemical fertilizer to lower an excessively high C:N would be useful.

### Central Processing (MRF)

Inasmuch as yard waste delivered to MRFs is likely to be contaminated with dirt and ferrous material, it usually (but not necessarily) is subjected to the processing indicated by the flow diagram in Figure 5.7. Size reduction (shredding) renders the waste easier to handle and brings its particle size distribution within that prescribed by product specifications. The size-reduced material is then screened and exposed to magnetic separation to remove materials that might interfere with certain uses of the product. For example, if the woody fraction is to be used as a fuel, it would be important that dirt, glass, and ferrous contamination be minimized. Removal of glass would be important if the processed waste is to be used as a compost feedstock. Equipment used in the processing (e.g., shredders, screens, magnetic separators, air classifiers) is described in Chapter 4.

Composting of yard waste processed for that purpose may be carried out at the MRF site. Alternatively, the processed waste may be transferred and composted at another site. Generally, the principles, methods, and technology are those described in Chapter 7.

Certain types of components may become problems when used as a compost feedstock, unless appropriate remedial measures are taken. Among the more important of such components are woody trimmings and large branches (either intact or size reduced), grass clippings, and fallen leaves (cf. section, "Components"). Probably the best approach with woody materials is to use them as a fuel or as bulking material for composting wastes that require bulking (e.g., sewage sludge). Problems with grass clippings come from the tendency of the clippings to form mats, which in turn become anaerobic. The matting problem can be avoided by thoroughly mixing the clippings with other components. In fact, if properly handled, grass clippings enhance the compost process, because they decompose readily and usually have an appreciable nitrogen content. Problems associated with odors due to the grass clippings generally are due to the decomposition that takes place during storage.

The problem with composting fallen leaves has several facets, among the more important of which are these three: (1) the greater part of the annual output, which can be sizeable, takes place within a 2- or 3-week period in autumn; (2) the C:N of the leaves is sufficiently high to materially slow the composting process (cf. Table 5.7); and (3) as long as they remain intact, some individual leaves are highly resistant to microbial attack. The leaf problem is not a serious one if sufficient space is available to permit a leisurely rate of composting, i.e., time is not of the essence and the dedicated land area is not required for other purposes.

Although nothing can be done about suddenness of the autumnal influx, the space requirements can be lessened by accelerating the compost process. The process can be accelerated significantly by size reducing or fragmenting the leaves, lowering the high C:N, and optimizing the operational and environmental conditions that are described in Chapter 7. Size reduction not only facilitates microbial access to the leaves, it also upgrades the final product in terms of appearance and handling. The C:N can be lowered by adding nitrogen either as a highly nitrogenous waste or as an agricultural chemical fertilizer (preferably urea). The addition of chemical nitrogen source must be carefully evaluated and justified since it can appreciably increase the operational cost of the facility.

The beneficial effect of lowering the high C:N of the leaves by adding chemical nitrogen has been demonstrated in studies conducted by the authors of this book. In one study, the initial C:N of the leaf feedstock were in the 50 to 80:1 range. The range was lowered to 19 to 26:1 through the addition of urea pellets. The addition was promptly followed by a substantial rise in temperature (i.e., from ambient to highs of 130 to 140°F), that could only be attributable to the improvement in C:N level; in as much as in the 2 months prior to the addition, temperatures had remained almost at ambient levels (40 to 70°F).

In the absence of unusual circumstances, the turned open windrow technology would be satisfactory. An alternative is the use of forced air (static pile) windrow method. With either approach, the windrows should be enclosed during the active stage of the compost process by a structure to shield them from rain and snow as well as from the very low temperatures characteristic of winter in some regions. In addition, the enclosure allows for proper control of dust and odors. In regions where relatively milder climates prevail, a roofed area would be sufficient. If time and space are not critical, protection with a tarpaulin would be sufficient, particularly during the rainy season. If for some reason or reasons, in-vessel technology is indicated, a simple form of the bin system would be suitable.

### *Trace Element and Pesticide Concentrations in Composted Yard Waste*

Among the constituents of yard waste and composted yard waste that pose a potential hazard to health and to quality and safety of our soil, water, and air resources are nitrogen, metals, pesticide residues, organic breakdown products, and other toxic organic compounds. Although general information on hazardous constituents in yard waste is becoming increasingly available in the literature, data regarding specific hazardous substances in yard waste and the environmental fate of those substances continue to be scarce. Similarly, much work remains to be done regarding the identity and fate of the intermediate breakdown products produced as the composting process proceeds from the raw yard waste to the finished compost.

Because of the many unknowns, prudence dictates that compost leachates, runoff, surface water, and groundwater beneath and adjacent to compost facilities should be closely monitored. In short, compost leachate and runoff water from the facility should not be allowed to reach surface and groundwaters.

## Food Wastes

### *Definition, Public Health, and Environmental Aspects*

Although the term is almost self-explanatory, it is well to attempt to trace the gradual development and application of the term over the past three or four decades. Originally, the term ''food waste'' referred to the biodegradable (putrescible) waste generated in the preparation and consumption of food and remaining after consumption (i.e., ''kitchen'' and restaurant wastes). With the passage of time, the term was broadened to include all discarded comestibles (e.g., spoiled or partially eaten fruit, stale bakery goods, etc.) and the vegetable trimmings generated in produce markets. ''Kitchen'' and restaurant food waste was the part of MSW that originally (pre-1970s) was known as the ''garbage fraction.'' In those days, the garbage fraction was collected separately from the other wastes (usually termed ''rubbish'') and was ''recycled'' as a feedstuff for swine. This practice, i.e., separate collection and hog feeding, gradually disappeared in the late 1960s and was replaced by ''combined'' (mixed garbage and rubbish) collection and disposal by way of landfill. Combined collection and disposal remained in favor until the 1980s, i.e., until the urgent need for resource recycling and landfill minimization began to be recognized.

A relatively recent development is the expanding advocacy for combining yard waste and food waste disposal — usually through composting. The concept has much in its favor. Food waste decomposes readily and may even enhance the compostability of yard wastes, especially of shrub and tree trimmings and leaves, by serving as a readily available energy source and to a limited extent as a nitrogen source for the microbial populations.

Unfortunately, putting the concept into practice could be encumbered by certain difficulties that could have unfavorable repercussions, some of which could be serious. An important difficulty is reconciling the seasonal variation in yard waste production with the year-round uniformity of food waste production. Because of this difference, yard waste would be in short supply during slack growing seasons (e.g., rainy season, later autumn and winter). However, the dearth of yard waste would be a problem only if the composition of the food waste were such that a bulking agent would be needed, in as much as yard waste can be an excellent bulking agent. With respect to repercussions, if mixed with food waste, yard waste would become a part of a mixture that would have all of the objectionable aesthetic, health, and environmental impacts usually associated with raw food waste (the garbage fraction of MSW). Objectionable impacts usually associated with food waste are intrusively visual and olfactory in nature. Among the objectionable health-related impacts is the attraction had by food waste for flies of all types (e.g., common house flies, fruit flies, "blow flies") and rodents, particularly rats. Moreover, food waste serves as nutrient source for the creatures as well as providing shelter for them.

### *Storage and Collection*

Because of the aforementioned, many unpleasant possibilities, the isolation of food waste from the environment during storage and collection must be as complete as possible. The high degree of isolation would necessitate the use of tight lidded, leakproof containers for storage. Moreover, because from 90 to 95% of the raw ("fresh") food waste mass is water, its density approaches that of water (1 gal water weighs about 8.3 lb). Therefore, the capacity of the storage container should not be larger than 10 to 12 gal if it is to be handled manually. If the food waste is to be transported to a central facility, the transport vehicle must be designed to ensure complete isolation (containment) of the food waste. For obvious reasons, storage time at the residence and at the central facility must be as short as is feasible. Good housekeeping practice must be maintained at the storage and the treatment sites.

Inasmuch as the aforementioned unpleasant possibilities and corresponding storage and collection requirements are not or are only minimally applicable to yard waste, practicality dictates that the storage and collection of yard waste be separate from those of food waste, and therefore that mixing of the two wastes should not take place until immediately prior to treatment (e.g., composting).

## Treatment (Composting) of Mixture of Yard and Food Wastes

### *Methodology*

The general composting methodology described for backyard and central facility composting of yard waste is almost the same as that for composting a mixture of yard and food wastes. A principal difference is that consequences of operational shortcomings and lapses are more severe in the composting of food-yard waste mixtures. In other words, the permissible latitude of departure from optimum feedstock composition and operational parameters is broader when yard waste alone is concerned. Additionally, the precautionary measures applicable to composting yard waste-food waste mixtures are more numerous and critical.

### *Residential (Backyard)*

A serious difficulty arises from the necessarily batch nature of backyard composting. Because of the batch nature, time elapses until sufficient waste to permit the imposition of operational procedures (particularly, turning) has been accumulated. Prior to the accumulation of sufficient material, diffusion of odors and access by pests could be minimized and perhaps prevented by covering the material with a tarpaulin underlain by a fine-mesh screen. Incidentally, the floor and sides of the bin should be constructed of "rat-proof" material (e.g., asphalt, concrete, or durable hardware cloth. However, the mesh size of standard hardware cloth is too large to prevent egress of insect and fly larvae from the bottom of mass of material.

### *Central Facility*

The problems associated with the batch type of composting perhaps could be reduced by resorting to a modified continuous type of composting. With the windrow method of composting (turned or static), continuity could be attained by adding material to one end of the windrow while more or less simultaneously removing an equivalent volume from the other end. In-vessel composting could be made continuous, although there would be some short-circuiting of unfinished material. Regardless of batch or continuous, raw and composting material both should be made inaccessible to flies and animals, especially rats.

Storage facilities for the raw food waste and for the mixed yard waste-food waste feedstock must be well constructed and maintained. These essential measures can best be attained by enclosing storage facilities and the active composting stage in a suitably designed structure.

## REFERENCES

1. Savage, G.M. and L.F. Diaz, "Processing of Solid Waste for Material Recovery," paper presented at the American Society of Mechanical Engineers, 14th National Waste Processing Conf., Long Beach, CA, June (1990).
2. Savage, G.M., "Design of Materials Recovery Facility (MRFs)," in Proc. of the First U.S. Conf. on Municipal Solid Waste Management — Solutions for the 90s, Vol. II, sponsored by the U.S. Environmental Protection Agency, Washington, DC, June (1990).
3. Sills, M.J. and R.N. Carrow, "Turfgrass growth, N Use, and Water Use under Soil Compaction and N Fertilization," *Agron. J.,* 75:488–492 (1984).
4. Rosen, C.J., N, Schumacher, R. Mugas, and S. Proudfoot, "Composting and Mulching: A Guide to Managing Yard Waste," Minnesota Extension Service, AG-FO-3296 (1982).
5. Epstein, E., and P.L. Engel, "Compost Products and Their Uses," *Solid Waste Power,* 5(4):44–53 (August 1991).
6. Brownio, A.H., *Geochemistry* (Engelwood Cliffs, NJ: Prentice-Hall, 1979).
7. "Portland Area Compost Market Study," Final Report No. 1189-3, prepared by CalRecovery, Inc. for the Metropolitan Service District, Portland, OR (October 1988).

8. McFarland, J.M., ''California Municipal Solid Waste Management: A Case Study in Public Enterprise,'' Doctoral thesis, submitted to Dept. Economics, University of California, Berkeley (1972).
9. Solbraa, K., ''An Analysis of Compost Starters Used on Spruce Bark,'' *BioCycle*, 25(2):46,47 (March 1984).
10. Golueke, C.G. and L.F. Diaz, '' 'Starters — Inoculums and Enzymes,' paper presented at *BioCycle* West Coast Conference '89, March 1989,'' *BioCycle*, 30(4):53–57 (April 1989).
11. ''Material Recovery Facilities for Municipal Solid Waste,'' Report No. EPA 625/6-91/031, prepared by PEER Consultants, P.C. and CalRecovery, Inc. for the U.S. Environmental Protection Agency, Office of Research and Development, Washington, DC, September (1991).

# CHAPTER 6

# Use of Organic Matter as a Soil Amendment

## INTRODUCTION

Although organic wastes contain plant nutrients which when incorporated into the soil increase its fertility, in most countries the nitrogen (N), phosphorus (P), and potassium (K) content of organic waste products usually is insufficient for the products to be legally classified as fertilizers. For example, to legally qualify as a fertilizers in the U.S., the combined NPK content would have to be at least 6%. However, the major utility of organic wastes in agriculture is not their NPK content, but rather their beneficial effect on the structure and other fertility characteristics of the soil.[1] Nevertheless, despite being secondary, the fertilizer aspect should not be overlooked.

The form in which the fertilizer elements, namely nitrogen, phosphorus, and potassium, occur in organic wastes has an important bearing on the application of the wastes. As would be expected, the three elements occur almost entirely in the form of organic molecules in unprocessed (raw) wastes, and to a lesser extent in processed wastes. In unprocessed wastes, the elements are bound mostly in plant and animal cellular protoplasm. As time goes on, the elements are gradually transformed into microbial protoplasm and into intermediate and final breakdown products, all of which adds up to decomposition. Ammonium-nitrogen is one of the more important of the breakdown products. The practical significance of fertilizer elements being bound in organic molecules is that only a part of the three nutrients becomes available to a crop immediately upon their introduction into the soil as an organic waste. Thus, it is estimated that only about 30 to 35% of the nitrogen in organic material such as compost is available for utilization by crop plants during the first year after application. However, in succeeding years, the nitrogen together with the other elements eventually become available.[2]

Raw organic wastes either can be used directly or they can be composted and then applied to the soil. The direct application of raw wastes may be preceded by some preparation, such as dehydration or shredding. Each method, i.e., apply directly or compost and then apply, has advantages and disadvantages, both of which become apparent in the sections that follow.

Organic wastes suitable for use as a fertilizer fall into three broad groups: garden and crop debris, animal and human wastes,* and food processing wastes (e.g., cannery wastes).

## USE OF RAW ORGANIC WASTES

### Preparation

The type and extent of preparation of the wastes depend upon at least four factors: (1) the physical characteristics of the waste; (2) whether or not the waste can be incorporated into the soil in its present state; (3) the time lapse (storage time) between generation and application of the waste; and (4) the distance to the site of use. The second factor is an extension of the first factor. Population density and proximity to the generation and application sites also are important factors.

### Physical Characteristics and Incorporation into Soil

Of the physical characteristics, particle size is among the decisive factors that determine the feasibility of direct application. The maximum permissible particle size is the largest size that allows manipulation by the machine and by the laborer who spreads it on the land. It is the size beyond which the waste can not be completely covered with soil or would require an inordinately deep incorporation into the soil to do so. The dimensions of the maximum particle size probably can be larger if the material is to be incorporated manually, because a machine has an inflexibly uniform depth tolerance, whereas a laborer can adjust tilling to suit the size of the particles — within reason, of course.

The molecular composition of the waste also has a bearing on maximum permissible dimensions. A material difficult to decompose (refractory) should be reduced to a smaller particle size than should be a fragile (e.g., dry leaf) or easily decomposed substance. For example, woody material should have a maximum particle size of 0.5 to 1 in., whereas green vegetable trimmings can be buried without undergoing any size reduction. The latter applies to green crop debris. Stalks of plants that are 3 ft or more in length should be chopped into about 12-in. sections — mainly for convenience of handling.

If the paper fraction is removed from the raw MSW, the remaining highly organic residue need not be size reduced — particularly in developing countries. If an appreciable amount of paper is present, the paper should be shredded to a maximum particle size of about 5 in.

Because of cost implications, it is advisable that size reduction be avoided whenever conditions permit the omission. Generally, if the direct utilization of

* Because of grave hazard to public health, human excreta must be treated in a manner such that pathogens and parasites are destroyed prior to making any use of the excreta.

raw waste is seriously considered, it is because a more sophisticated approach is not readily attainable within the financial and economic constraints existing in the particular area.

Dehydration is useful because it can be done advantageously. Dehydration is indicated in situations in which storage time and distance of transport are significantly long. It also might be required if the generation, collection, and disposal sites are in relatively densely populated areas.

Dehydration also is useful because it can bring about a more or less complete suspension of microbial activity and hence prevent the production of breakdown products of an objectionably odorous nature. Microbial activity comes to a complete stop when the moisture content drops to approximately 12% and declines rapidly as that level is approached. Moreover, dehydration renders the material less attractive and less suitable as a foodstuff and shelter for most vectors, but not necessarily for rodents. Additional advantages are an increase in ease of handling and a minimization of drainage during storage and transport. Issues involving odors and vectors become especially critical, even in areas that are only moderately densely populated. Ease of transport is advanced and the production of leachate is lessened because of the drop in weight and volume of the raw wastes that accompanies dehydration. This is true because dehydration can reduce by as much as 70 to 75% of the 85 to 95% moisture found in most living matter.

Dehydration should be limited to that attained through air drying, preferably by exposure to the sun. Spreading on an exposed surface or placing on racks are two ways of bringing about dehydration. The cost of accomplishing dehydration of a waste in any manner other than the preceding two becomes prohibitive even in industrialized countries.

Finally, an approach first tried in Odessa, TX, and later in Israel should be mentioned.[3] The attempt in Texas proved to be unsuccessful, whereas the Israeli attempt apparently had a successful outcome. In the Israeli attempt, the organic garbage (i.e., wastes associated with the preparation of food) was spread on a barren spot 12 mi from Beersheba; 8 tons of the waste were spread over 0.1 acre of desert wasteland. The waste was sowed with seeds of a high protein animal feedstuff and a white wheat. Seeds and waste were covered with sand. The plot was not watered. The yield from the 0.1-acre plot was 4 tons of animal feed. According to the investigators, substantial yields were obtained over the subsequent 10 years without resorting to watering or further reseeding.

### Advantages

An important advantage in the direct application of raw waste to the land is the substantial monetary saving that can be achieved by omitting major processing steps. The savings are due to the elimination of the need to acquire and operate processing equipment. A second advantage is a probable reduction of the loss of nutrients that often is a direct result of processing (e.g., loss of ammonia-nitrogen during composting). If the material can be used immediately, land area requirements are reduced drastically. Moreover, allowing for the constraints and limitations described in the following few paragraphs, the benefits that result from the incorporation of organic matter in soil can be attained with a smaller expenditure of effort, time, and money.

## Limitations or Constraints

Unfortunately, the advantages to be gained from the application of raw organic wastes are strongly tempered by some limitations. These limitations apply to crop production, maintenance of a good quality environment, and safeguarding the public health.

### *Crop Production*

Studies in Europe indicate that in terms of promoting plant growth, composted waste significantly surpasses raw organic waste. Moreover, common gardening and farming experience demonstrates that unless precautionary measures are taken, the application of a fresh ("green") manure can lead to very deleterious consequences in crop production. The severity of the damage depends upon the animal source of the manure. For example, damage from applying fresh cow manure, though significant, is much less severe than that from applying fresh chicken or swine manure. The consequences from the use of unprocessed crop debris and urban refuse are not quite as drastic.

A severe form of the damage is the so-called "burning" that may follow the application of a fresh raw waste. It resembles the "burning" that is the immediate consequence of an over application of chemical fertilizer. The burning takes the form of necrotic zones on the edges of the leaves. In less severe cases, the damage is manifested as a general stunting of plant growth.

A commonly accepted explanation is that the damage is brought about by the release of toxic concentrations of ammonia in the root zone. If the concentration is sufficiently great, the roots are killed. Another possibility is the formation of toxic breakdown products. Researchers who have observed a less positive response to raw waste or less than fully matured compost in comparison to that of properly matured compost do not have a complete explanation for the difference.

The unfavorable impact on plants can be lightened either by spreading the material on the field in the autumn or by interposing a 4- to 8-in. layer of soil between the raw waste and the root zone. The thickness of the layer of soil can be reduced on the assumption that the waste will be almost decomposed before the seeds germinate. In such a case, the root systems of the seedlings will have become extensive enough to have penetrated the layer of wastes. By spreading the wastes in the autumn, sufficient time is allowed for decomposition to take place before the subsequent planting season. In regions characterized by winter snows, there is the danger of eutrophic elements being transported to receiving waters by way of the melting snow. Pollution of receiving waters is also possible in regions characterized by heavy rainfall during the period when the fields lie fallow.

### *Public Health Hazard and Environmental Degradation*

The limitations described in the following paragraphs apply only to raw wastes that are highly putrescible or are animal wastes (manures). The stability of many types of crops is sufficient to place them outside of this category.

Unless suitably protected during storage and transport, raw organic wastes serve not only as attractants, but also as sources of nutrient and as shelters for rodents and vectors. Rodents are attracted by wastes of plant origin and rarely, if at all, by manures. For example, in the U.S. it is well known that an open dump containing municipal refuse can serve as a focal point for the spread of rat populations over a circular area of approximately 5 mi in radius. The spreading area of flies is about the same. The entire reproductive cycle of flies can and does take place in decaying organic matter. Major deterrents to their development are a low moisture content and either very low or very high temperatures.

Two other disadvantages are aesthetic in nature. One is the production of objectionable odors. Under certain conditions, the foul odor emission may be very intense. The second aesthetic affront is the unsightliness of raw wastes.

Environmental and public health problems can be solved by preventing access by undesirable microorganisms and macroorganisms during storage and transport. In the field, they can be controlled by incorporating the wastes into the soil as soon as possible.

## Economics

If processing is not required, the major cost prior to application is that of transportation. Transportation costs can constitute the deciding factor in the decision to use or not to use a raw waste. An important point to bear in mind is that with the increasingly higher costs of fuel and the corresponding expanding rate of inflation, the economically permissible longest distance of transport is soon reached. The unit cost for truck transport in the U.S. and in other countries has been doubled and even tripled since the late 1960s.

If the material is to be stored, then cost of building a suitable shelter must be taken into consideration. However, the shelter need not be elaborate. Access by rodents and flies to stored waste can be prevented by covering it with screening, although that would do nothing with respect to unsightliness and odor emission.

Health authorities in the U.S. have come to the conclusion that the only satisfactory way of storing beef and dairy cattle manures is to place the manures in a concrete tank. By so doing, runoff and leachate formation are prevented and flies do not have access to the wastes.

## USE OF COMPOSTED WASTE

### Description of Product

Properly composted material has the desirable characteristics of humus. The term "humus" was defined by Waksman[1] as being a complex aggregate of amorphous substances resulting from the microbiological activity that takes place in the breakdown of plant and animal residues. In terms of chemical makeup, it is a heterogeneous mixture of substances that includes a variety of compounds synthesized by the microbial populations, of the complexes resulting from the microbial decomposition, and of materials that resist further breakdown. Thus, its principal constituents are derivatives of lignins, proteins, certain hemicelluloses, and celluloses.

As a humus, compost is not biochemically static. Consequently, under appropriate environmental conditions, it will be further decomposed by microbes and to some extent by higher forms of life (e.g., earthworms and a number of insects) until eventually it is oxidized to mineral salts, carbon dioxide, and water. Because of its high humus content, compost has an ample capacity for base exchange, with consequent swelling.

The composting material takes on an earthy odor towards the end of the composting phase. At this time, fungi and actinomycetes usually become visible. The earthy odor mingled with slightly musty overtones continues to be a characteristic of the product long after the processing phase has been completed.

Directly after the composting phase is completed, the appearance of the finished product strongly depends upon the physical characteristics of the waste that was composted. The dependence is most pronounced with respect to particle size, which probably is responsible for any overall similarity in the appearance of the compost product to that of the raw material. However, regardless of origin, the product is deep brown to dark gray in color. Reference is made to appearance directly after the composting phase, because unless the moisture content of the material is 15% or less, decomposition continues and therefore the appearance also continues to change. The appearance directly after the composting phase is the one it has when it is ready for distribution to users. Regardless of original material, after prolonged storage it eventually approaches the consistency of a fine dust. Because the eventual change in appearance also betokens a radical change in properties, all descriptions in this section refer to the compost product directly after it has been produced, i.e., has been satisfactorily stabilized. If a municipal refuse contains a sizeable paper fraction, the compost product may have bits of recognizable paper. Similarly, straw, wood, or other resistant material may be recognizable as such in the finished product. Despite the similarity in appearance, the nature of the material has changed. For example, particles are much more brittle than they were prior to composting. The change is manifested as a drastic increase in ease of shredding.

Product appearance can be materially enhanced through screening to produce a compost that has a relatively uniform and small particle size. Because glass shards and pieces of plastic detract from the appearance and lessen the utility of the product, they should be removed to the extent possible.

As stated earlier, the nitrogen, phosphorus, and potassium (NPK) content of the usual compost product is not sufficiently high to legally justify the designation, "fertilizer" in the U.S. The NPK of the product reflects that of the raw material. Thus, if the raw waste has a substantial NPK, so will the compost product; whereas a low NPK is followed by a comparably low NPK in the product.

Examples of nitrogen concentration (dry wt basis) of municipal refuse typically encountered are as follows: U.S.A., 0.5 to 1.5%; São Paulo, Brazil, approximately 2% to 4%; and Mexico City, approximately 2 to 4%. The nitrogen content of composted cattle manure ranges from 0.5 to 2.5%. The nitrogen content of composted fowl manure, pig manure, and sheep manure reflect the substantial nitrogen concentrations of the raw manures. The phosphorus content of composts generally is only slightly lower than that of the nitrogen content. When municipal refuse is composted, the potassium content generally is much lower, unless wood ash were present. The low potassium content is not a serious nutritional deficiency in areas where the soil is alkaline.

The carbon-nitrogen ratio (C:N) declines during the compost process. If the process is carried out properly, the C:N of the product does not exceed 20:1. The gradual decline is due to the loss of C as carbon dioxide formed in metabolism.

The 20:1 is a critical ratio in terms of crop production. If the C:N is higher than about 20:1, a strong possibility of a nitrogen shortage for the crop plants arises. The shortage is ultimately a result of the metabolic and synthetic activities of the microflora and of the nutritional requirements of the crop plants. As is characteristic of all living cells, metabolism and synthesis of new protoplasm occur simultaneously when sufficient nitrogen is available. If the supply of nitrogen is insufficient to fulfill the demands of all organisms present in the soil, the organisms compete with each other for the available nitrogen. Organisms most efficient in nitrogen assimilation take up most of the available nitrogen — at the expense of the less efficient organisms. The latter then show symptoms of nitrogen deprivation. Because bacteria are more efficient than crop plants in assimilating nitrogen, the latter suffer a shortage of nitrogen, which is manifested by chlorosis and a general stunting of plant growth. An excessively high C:N can be lowered by applying a nitrogen fertilizer to the soil simultaneously with the compost.

A C:N that is too low exposes crops to the "burning" that comes from the application of "green" (insufficiently aged) manure — and for the same reasons. The danger level probably begins at a C:N lower than about 10:1.

The variations in visual and nutritive characteristics of composts make it strongly advisable to grade the product, either as a single mass or as subdivided into portions separated on the basis of differences in quality. In fact, the most effective use of a given compost product presupposes grading. The utility of grading stems partly from the fact that applications differ among themselves with respect to the quality of the compost required. For example, reclamation of land despoiled by strip mining can be satisfactorily accomplished with a relatively low grade of compost; whereas, a very high-grade product would be needed in vegetable farming.

Grading can be based on NPK content; particle size and uniformity; amount of contaminants such as glass and plastics; freedom from pathogens and toxic substances; and possibly, degree of "maturity." Certainly, a product that is free of contaminants, that is safe in terms of public health, that has a uniformly small particle size, and that has an NPK approaching that legally expected of a fertilizer (about 6%) would be classified as a top-grade compost (Grade 1). It would be suitable for the more demanding applications, as for example, production of vegetables destined for human consumption either before or after having been cooked.

The following two lower quality grades and their specifications are merely suggested. Other standards could be established to meet situations specific to the area in which the compost operation takes place. The second grade (Grade 2) would be a compost product that has a maximum particle size larger (1 to 2 in.) than that of the grade 1 product, includes some glass and plastic bits, has a maximum C:N of about 20:1, and is free of pathogens. It would be suitable for use in orchards and in raising field crops. The grade 3 product would include composted material that does not meet the specifications for Grades 1 and 2. It could be used in land reclamation schemes. For further information on grading, the reader is advised to consult References 4 and 5.

## Method of Applying Compost

Regardless of whether it has been designed for mechanical or for manual application, equipment used for applying raw organic waste can also be used for applying compost. The principal difference in use in the field is that the compost product can be allowed to come in direct contact with the plants. This latitude removes the need to interpose a layer of soil between root zone and the compost. Moreover, the compost can be used as a mulch without constituting a nuisance.

## Loading

The balance between the amount of nutrients (NPK) consumed by the crop and the amount added by way of the compost is a key factor in the determination of loading. Of the three elements, nitrogen is of the greatest concern. Accordingly, the maximum amount of compost that can be applied without giving rise to a problem is that which contains precisely the amount of nitrogen used by the crop during the growing season. For example, if corn is the crop to be raised, then about 200 lb of nitrogen would be required per acre. If the compost to be added to the field has a nitrogen concentration of 2% by weight, the maximum

**Table 6.1 Quantity of Nitrogen Released Annually per Ton of Compost Incorporated in the Soil**

| Organic N in Compost (%) | Amount of Nitrogen Released (lb/Ton Compost Added) | | |
|---|---|---|---|
| | Year-1 | Year-2 | Year-3 |
| 2.0 | 14.0 | 14.0 | 12.0 |
| 2.5 | 17.5 | 17.5 | 15.0 |
| 3.0 | 21.0 | 21.0 | 18.0 |
| 3.5 | 24.5 | 24.5 | 21.0 |
| 4.0 | 28.0 | 28.0 | 24.0 |
| 4.5 | 31.5 | 31.5 | 27.0 |
| 5.0 | 35.0 | 35.0 | 30.0 |

**Table 6.2 NPK Requirements by Various Crops**

| Crop | Yield (bu) | Phosphorus (lb/acre) | Nitrogen (lb/acre) | Potassium (lb/acre) |
|---|---|---|---|---|
| Corn | 168 | 35 | 184 | 177 |
| | 202 | 44 | 239 | 200 |
| Corn silage | 36 | 35 | 200 | 202 |
| Soybeans | 56 | 21 | 257 | 100 |
| | 67 | 29 | 335 | 119 |
| Grain sorghum | | 40 | 250 | 166 |
| Wheat | 67 | 22 | 125 | 90 |
| | 90 | 24 | 185 | 134 |
| Oats | 112 | 24 | 150 | 125 |
| Barley | 112 | 24 | 150 | 125 |
| Alfalfa | 9 | 35 | 450 | 397 |
| Orchard grass | 7 | 44 | 300 | 310 |
| Brome grass | 5.6 | 29 | 166 | 210 |
| Tall fescue | 4 | 29 | 134 | 153 |
| Bluegrass | 3.4 | 24 | 200 | 149 |

permissible loading theoretically would be about 5 tons. However, because only about 30 to 35% of the compost nitrogen is available to the crop in year 1, the loading could be about 14 to 17 tons. A point to keep in mind in determining loading in the immediately succeeding years is that the remaining nitrogen eventually becomes available to the plants, i.e., 30 to 35% in year 2, and the remainder in year 3. An indication of actual amounts released each year is given by the data in Table 6.1. The reason for maintaining balance between nitrogen addition and nitrogen consumption is that soil bacteria oxidize excess nitrogen to nitrate, i.e., mineralize the nitrogen. Because the nitrate is soluble, it can be leached to the groundwater and thereby deteriorate the water. Amounts of NPK required by various crops are listed in Table 6.2.

If the compost contains materials toxic to plants or to animals and to people who consume the plants and meat and products of animals, then the maximum permissible loading is determined by the concentration of the objectionable substances in the compost, times the number of tons at which the upper permissible loading would be reached. Of course, the maximum amount that can be applied without damage is the difference between the amount at which damage occurs and that already in the soil. The number of tons per acre is easily estimated according to the equation

$$x = \frac{y}{z}$$

in which x is total amount of compost per unit of area (acre), y is the permissible maximum amount (lb) of the objectionable substance per acre, minus the amount already in the soil, and z is the concentration of the substance per unit mass of compost (lb/ton).

Factors that determine the concentration in the soil at which a toxic element becomes inhibitory to a crop plant or endangers the health of a consumer of the crop are (1) pH of the soil; (2) organic matter in the soil; (3) degree of aeration of the soil; (4) structure and ion exchange capacity of the soil; and (4) amount of uptake of the element by the plant. The mechanism through which the factors exert their influence on the soil is by way of their effect, combined and individually, on the solubility of the substance in question. Solubility is the ultimate factor because (1) a plant can only assimilate soluble substances; and (2) the toxic substances can exert their harmful effects only if they are assimilated by a plant. The exception is if the substance is physically destructive of plant tissue, e.g., ammonia on the root hairs of plants.

A substance becomes hazardous to humans only when the concentration of the substance in the edible portion of the plant is at a level toxic to the consumer. Generally, the heavy metals (toxic to plants and potentially so to humans) are insoluble at pH levels higher than 7.0 (alkaline soils). The extent of immobilization or fixation by chelation, ion exchange, or by being rendered insoluble is magnified by increasing the amount of organic matter in the soil, by promoting aeration in the soil, and by ensuring a high exchange capacity in the soil. The addition of compost amplifies all three factors. Consequently, the upper permissible limits are raised through the application of compost to the soil. This fact should be taken into consideration in the determination of maximum permissible loadings. For a detailed discussion on the heavy metal problem, the book by Leeper[15] is an excellent reference.

In summary, a generally acceptable loading of compost is about 5 to 10 tons/acre/year; or from 10 to 15 tons/acre/3-year interval. If the 10 to 15 tons are all added in year 1, no further additions are applied in years 2 and 3. Exceptions in the form of reduced loadings would be indicated with composts made from one or all of the following: (1) sewage sludge from an industrial sector; (2) from a waste containing sizeable amounts of wood or coal ash; and (3) a material having an exceptionally heavy concentration of nitrogen.

### Advantageous Effects Exerted on the Soil by Compost

The advantages include those attributed to raw waste in the preceding section. One such advantage is enrichment of the NPK content of the soil. Because of the NPK in the compost, the amount of the three elements added in the form of a chemical fertilizer can be reduced in proportion to their concentration in the compost. Indeed, the amount of reduction in required dosage of chemical fertilizer is more than the simple difference between total amount required and that in the compost. The presence of compost (or any organic matter of plant or animal origin) increases the efficiency of chemical NPK utilization. This is done through the conversion of the fraction of the chemical fertilizer not used by the crop into microbial cellular material. NPK thus fixed (converted) is slowly released as microbes die off.

In the absence of organic matter, from 30 to 35% of the nitrogen, 20 to 30% of the phosphorus, and a lesser fraction of the potassium added as chemical fertilizer are leached beyond the root zone and thus are of no value to crop production.

Phosphorus is stored through the agency of organic acids synthesized in the metabolic breakdown of organic matter. The acids form a complex with inorganic phosphates in the soil and thereby render the phosphorus more readily available to higher plants.

In summary, phosphorus as well as nitrogen are, in effect, "stored" in a manner peculiar to humus in that precipitation of phosphorus by calcium is deterred, and nitrogen, as stated earlier, is converted into microbial protoplasm. The conversion of nitrogen into microbial protoplasm interferes with the mineralization (nitrification) of nitrogen that would otherwise be the fate of nitrogen introduced into the soil. Prevention of nitrification in turn protects the quality of groundwaters from being degraded. Not to be overlooked are the valuable trace elements contained in compost. Almost all trace element deficiencies in a soil can be eliminated through the application of compost.

A second major advantage due to the use of compost is the beneficial effect exerted on soil structure. Structure ranks with nutrient content in determining the productivity of a given soil. The improvement in soil structure results from the tendency of compost to bring about soil aggregation, which in turn imparts a crumblike texture to the soil, i.e., makes the soil friable. Friability is tied in with soil aeration and water holding capacity. The more friable a soil is, the greater is its capacity to retain water and the higher is its state of aeration. Since aeration and moisture are important factors in the development of root systems, plants grown in soils containing compost are characterized by well-developed root systems. The aggregation is accomplished through the agency of various cellulose esters formed in microbial metabolism.

Another aspect of the water conservation potential of compost is its use as a mulch. As such, the layer of compost serves as a physical barrier to evaporation.

Its usefulness as a mulch is further enhanced by the fact that the compost eventually can be incorporated into the soil and further its beneficial effects in that medium.

The beneficial effects with respect to chelating and fixing heavy metals were described in the section on loading.

An important benefit is an accompanying increase in the resistance of crops to many plant diseases. Whether or not the increased resistance is due to the more vigorous growth encouraged by the presence of compost or to an actual development of an immunological response in the plant itself is immaterial. The result is the same. For example, it has been demonstrated that in the presence of composted hardwood bark, normal growth of citrus trees can be obtained in soils heavily infested with nematodes.[6] It also has been shown that the growth of *Pythium* is suppressed through the use of compost produced from pine bark.[7]

The discussion of advantages is closed with this observation: the substitution of chemical fertilizers in whole or in part in agriculture is gaining increasingly favorable attention not only for family-scale farms, but also for large-scale farms (''megafarms''). The substitution has long-term ecological, sociological, and public health ramifications that will become more apparent as the conversion process from high- to low-input agriculture continues.[8]

Erosion has a significant negative impact on agricultural practices. A major consequence of erosion is the loss of plant nutrients. In the U.S., 1.54 million tons of N, 34 thousand tons of P, and 1.27 million tons of K are lost each year through erosion. The major impact of soil deterioration can be quantified in terms of fertilizer, yields, and power or fuel requirements. The A horizon normally is the most productive section of soil profile. As illustrated in Figure 6.1, crop yields decrease as the A horizon is eroded. Soils vary in their response to fertilizer inputs as erosion progresses. The change is strikingly exemplified by the data in Table 6.3. Despite increased fertilizer inputs, yield generally decline as a soil shifts from one erosion phase to another. The corresponding decline in yield is exemplified by the data in Table 6.4. In a report published by the National Research Council,[9] it is pointed out that the average corn yield is reduced 4% for each 2.5 cm of topsoil removed (lost) from a base of 30 cm. The data in Table 6.5 further illustrate the effect. With respect to fuel requirements, erosion phase 1 soils require 22% more fuel for tillage when depleted to erosion phase 2. A 17% increase in fuel was needed to till erosion phase 2 soils after they had been depleted to erosion phase 3. The National Research Council report states that 75 L of oil are needed to offset the increase in energy required with a 220-kg/ha loss in soil.

An example of the total increase in costs due to soil depletion is given in the data in Table 6.6. Of course, the monetary loss that would occur would be functions of the monetary value of the crop and cost of fuel.

To the preceding adverse impacts had by erosion, most researchers add yet another impact, a loss of plant-available soil water capacity. The water-holding characteristics of the root zone are changed when topsoil is removed, because

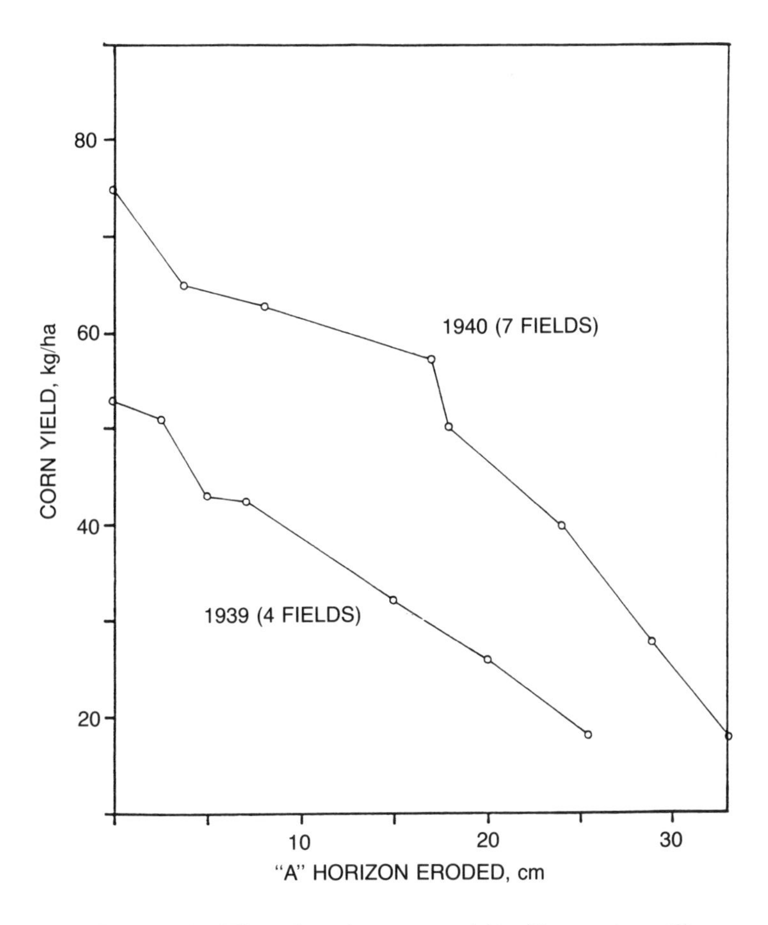


**Figure 6.1.** Effect of erosion on corn yields. (*Source:* Jenny.[12])

**Table 6.3 Increase in Fertilizer Needs (lb/Acre) as Soil is Depleted, Southern Iowa Rivers Basin**

| Change in | Corn | | | Soybeans | | Oats | | | Meadow | |
|---|---|---|---|---|---|---|---|---|---|---|
| Erosion Phase | N | $P_2O_5$ | $K_2O$ | $P_2O_5$ | $K_2O$ | N | $P_2O_5$ | $K_2O$ | $P_2O_5$ | $K_2O$ |
| 1–2 | 10 | 2 | 6 | 1 | 5 | 5 | 2 | 6 | 2 | 10 |
| 2–3 | 30 | 1 | 7 | 1 | 7 | 15 | 1 | 7 | 1 | 12 |

*Source:* Unpublished data, Iowa Agricultural Experiment Station, Ames.

**Table 6.4 Reduction in Yields as Soil is Depleted, Southern Iowa Rivers Basin**

| | Reduction in Yield per Acre | | | |
|---|---|---|---|---|
| Change in Erosion Phase | Corn (bu) | Soybeans (bu) | Oats (bu) | Hay (tons) |
| 1–2 | 16 | 5 | 9 | 0.6 |
| 2–3 | 7 | 3 | 4 | 0.5 |

**Table 6.5 Relation Between Topsoil Depth and Yield of Corn Drawn from Selected Studies (Compared with Standard Plots of 12-in. Topsoil Depth, Corn Yields were Decreased by the Amounts Shown When Corn was Grown in Soils of the Depth Indicated)**

| Topsoil Depth (inches) | Yield (bu/acre) | | |
|---|---|---|---|
| | Range | Average | Decrease |
| 0 to 2 | 25 to 56 | 36.2 | 10.8 |
| 2 to 4 | 28 to 69 | 47.0 | 9.3 |
| 4 to 6 | 39 to 83 | 56.3 | 8.4 |
| 6 to 8 | 49 to 97 | 64.7 | 4.3 |
| 8 to 10 | 50 to 102 | 69.0 | 5.3 |
| 10 to 12 | 50 to 125 | 74.3 | |

*Source:* Pimentel et al.[10]

**Table 6.6 Specified Annual Costs of Soil Depletion,[a] Southern Iowa Rivers Basin**

| Item | Million Dollars in the Year | |
|---|---|---|
| | 2000 | 2020 |
| Fertilizer | 2.0 | 2.6 |
| Yields | 4.3 | 8.1 |
| Fuel | 0.4 | 0.7 |
| Total | 6.7 | 11.4 |

[a] Assuming 1974 conditions throughout the period to 2020.

topsoil usually has a higher plant-available water capacity than has subsoil. Although for a time, the adverse effects of erosion on yield can be counteracted by optimizing water and nutrients, eventually erosion will raise the costs of optimization until they become prohibitive and make it impossible to maintain production levels.

Various methods can be used to control or minimize erosion. Contour planting is one method. The problem is that contour farming can result in a 5 to 7% increase in both farming time and fuel use. Crop rotations, where they can be done, can help conserve soil. A combination of contour planting and crop rotation provides more soil erosion control than either would individually.

Where the situation is such that contour planting and crop rotation methods are infeasible, an effective erosion control measure is to increase the organic matter of the soil. In one reported study, it was found that the addition of 35.5 tons of manure per hectare to corn land in Iowa (U.S.) that had a slope of 9% resulted in an average annual soil erosion of only 10.5 tons/ha, as compared to an average of 49.3 tons/ha on land that received no manure. In a report issued by the World Bank,[11] reference is made to the fact that the addition of 400 tons of compost/hectare resulted in a 95% reduction in soil erosion. In an FAO report,[13] a study is cited in which, with the slope at 58%, the loss in soil was 30% less (dry weight) when compost was added at 89 tons/ha and 96% less when compost was added at 178 tons/ha.

## Limitations

Because limitations with respect to crop production were covered in large part in the section on loading, the discussion in this section is devoted to (1) limitations pertaining to the environment and to public health not covered earlier, and (2) limitations pertaining to the transport and to the mechanics of applying the compost on the field.

### *Environment and Public Health*

The major potentially unfavorable impacts upon the environment from the use of compost are those that result from overloading to the extent that the amount of nutrient exceeds that used by the crop. The consequences were described under ''Loading.''

Public health hazards arise when the product is composted human excrement, or a sludge derived from wastewaters of industrial origin, or a composted hazardous waste. The composted hazardous waste may contain toxic intermediates. Wastes from diseased animals could be ranked with human excrement as a zoonotic hazard to public health, but perhaps to a lesser degree. A hazard to public health exists (1) if the pathogens are not exterminated during or after the composting process; or (2) if part or all of the finished product is contaminated with a fresh waste that contains pathogens. Such contamination occurs only as a consequence of poor management.

The act of composting per se is not a guarantee that all pathogenic organisms have been killed and, therefore, that the product is free from such organisms. The common conviction to the contrary is a result of the fallacious belief that the high temperatures reached in a pile of composting material result in the destruction of all pathogens in the composting mass. The fact is that many factors combine to thwart a complete kill-off, and hence it serves no purpose merely to make up a table of the thermal deathpoints of a variety of pathogens.

An important factor responsible for an incomplete kill is the failure to expose all parts, however small, of a pile to a temperature lethal to pathogens. A small amount of insufficiently exposed infested or contaminated material can recontaminate an entire pile of compost. Moreover, the process of turning (described in the chapter on composting) results in a recontamination of the ''sterile'' section of a pile by the material not as yet exposed to high temperatures. In systems that involve no turning, sufficiently elevated temperatures may prevail throughout the pile; but even in them, areas of low (approaching ambient) temperatures may exist. Despite these limitations, the attainment of bactericidal temperatures does make a significant contribution to the overall destruction of pathogens.

The shortcomings of temperature are compensated by other factors which collectively require a passage of time ranging from 1 to 12 months, and in rare cases, longer, to bring about a relatively complete destruction of pathogens. The destruction need not be entirely completed; a reduction in number to a level at which they no longer are infective is sufficient.

Additional factors are destruction of nutritional source, competition with non-pathogenic microorganisms, antibiosis, and a less-than-favorable environment.

In as much as the normal habitat of pathogens is the human body, nutritional sources useful to pathogens are compounds found in the body. The fact is that only a relatively few compounds can be used by pathogens as a source of nutrients. These compounds are rapidly destroyed during composting. Because they are adapted to the protected environment provided by the bodies of their hosts, pathogens are poor competitors with the microbial populations indigenous to the compost feedstock. The normal habitat of the latter is the external environment; whereas pathogens generally do not proliferate in the external environment.

The usual presence of actinomycetes and fungi during composting ensures the production of a variety of antibiotics. For example, species of *Streptomyces* (actinomycete) and *Aspergillus* (fungus) appear in substantial numbers during composting. Therefore, all things considered, a year of storage should ensure the destruction of all but the most resistant of parasitic forms, microbial spores (e.g., anthrax), and pathogens (e.g., *Mycobacterium tuberculosis*). Because they all are very sensitive to ultraviolet radiation, even resistant forms can be eliminated by spreading the compost in a thin layer exposed to the sun for a week or two. The layer should be raked each day.[14]

Some concern exists regarding the public health significance of *Aspergillus fumigatus*, in as much as the fungus and its spores generally, but not invariably, occur in the compost product. However, it has been demonstrated that the fungus is of little if any danger even to the very sickly and elderly. At most, it could be an allergen for a susceptible individual. The fact is that the fungus is a saprophyte commonly found in decaying vegetable matter — e.g., fallen leaves.

### *Transport and Mechanics of Applying Compost*

Difficulties pertaining to transport and the mechanics of application stem almost entirely from the bulky nature and low density of the dry compost product. Obviously, the cost of transporting a low-density product is greater than that of transporting a high-density one. The volumetric capacity of a given transport vehicle for a low-density certainly is reached long before it is for a high-density (compacted) compost product. The bulkiness and low density of compost necessitate the use of bulky equipment to spread it on the field. Although this aspect may not necessarily be significant if the material is spread manually, the fact remains that more trips per unit mass of material are needed. However, these difficulties diminish to the level of minor importance when compared to the many benefits to be gained from the use of compost.

## SUMMARY

The preceding discussion of the benefits that can be obtained through the use of organic waste as a soil amendment and organic fertilizer, and of enhancement of the benefits through composting, logically is followed by a chapter devoted to supplying the information needed for efficiently carrying out the compost process.

## REFERENCES


1. Waksman, A.A., *Humus*, 2nd ed. (Baltimore: Williams & Wilkins 1938).
2. ''Second Interim Report of the Interdepartmental Committee on Utilization of Organic Wastes,'' *N.Z. Eng.*, 6 (6–12), November-December (1951).
3. Ragen, N., ''The Struggle for Israeli's Environment,'' *Hadassah Mag.*, November (1979).
4. ''Compost Classification/Quality Standards for the State of Washington,'' Final Report, prepared by Cal Recovery Systems, Inc., for the State of Washington Department of Ecology, September (1990).
5. Eggerth, L.E. and C.G. Golueke, ''Monitoring Compost Quality,'' presented at the BioCycle Southeast Conference '90 on Successful Recycling for Solid Waste and Sludge, Miami Beach, FL, October (1990).
6. Hoitink, H.A.J. and P.C. Fahy, ''Basis for the Control of Soil-Borne Plant Pathogens with Composts,'' *Annu. Rev. Phytopathol.*, 24:93–114 (1986).
7. Chen, W., H.A.D. Hoitink, and L.V. Madden, ''Microbial activity and biomass in container media for predicting suppressiveness to damping-off caused by *Pythium ultimum*,'' *Phytopathology*, 78:1447–1450 (1988).
8. Altieri, M., et al ''Low-input technology proves viable for limited-resource farmers in Salinas Valley,'' *Calif. Agric.*, 45(2):20–23 (1991).
9. Pimentel, D. *Issues and Studies 1981–1982*, National Research Council (Washington DC: National Academy Sciences Press, 1982), pp. 105–115.
10. Pimentel, D., E.C. Terhune, R. Dyson-Hudson, S. Rochereau, R. Samis, E.A. Smith, D. Denman, D. Reifschneider, and M. Shepard, ''Land Degradation: Effects on Land and Energy Resources,'' *Science*, 194(4261):149–155 October 8 (1976).
11. World Bank, ''Organic Fertilizers: Problems and Potential for Developing Countries,'' World Bank Fertilizer Study, Background Paper No. 4, January 1974, IFC-Office of the Economics Advisor (15 pp.).
12. Jenny, H. *The Soil Resource: Origin and Behavior* (New York: Springer-Verlag, 1980).
13. ''Potential Health Hazards and Legal Implications of Waste Recycling,'' New Feed Processes, FAO Animal Production and Health Paper 4, Food and Agricultural Organization, Rome, 1977.
14. Golueke, C.G., ''Epidemiological Aspects of Sewage Sludge Handling and Management,'' in Proc. Int. Symp. Land Application of Sewage Sludge, Tokyo, October (1982), pp. 165–214.
15. Leeper, G.W., *Managing the Heavy Metals on the Land* (New York: Marcel Dekker, 1978).

# CHAPTER 7

# Composting

## INTRODUCTION

Because composting is a biological decomposition process, it has the many advantages and the limitations normally characteristic of biological systems. In comparison with nonbiological systems, two of the advantages of biological systems are a generally lower equipment and operating outlay and fewer unfavorable impacts upon the quality of the environment. Disadvantages often attributed to biological systems are slow rate of processing and unpredictability. Although it may be true that rate of processing may be more or less slow in biological systems, the attribution of unpredictability is no more justified for a properly operated biological system than it would be for a nonbiological system. If conditions are known and the design capacity of the system is not exceeded, then the process will progress precisely as predicted.

## BASIC PRINCIPLES

An awareness and understanding of the basic principles of composting are essential to a rational evaluation of the technology, of the process, and its applications. Ignorance of or failure to give those principles the consideration due them can lead either to a selection of a system only remotely suitable for a given situation, or to one that is needlessly sophisticated, or to both. A needlessly sophisticated design is by that very fact excessively expensive. In fact, it is quite possible that the chosen system could be so completely out of line with the real situation as to be foredoomed to failure. Although applicable to every technological process, the aforementioned statements are especially apropos to composting. The reason is that the economic political feasibility of composting can be directly and indirectly imperiled by exaggerated claims made either by overly enthusiastic proponents of composting or by proponents of particular processes and equipment.

## Definition

Underlying our description and discussion of the basic principles of composting is the following engineering definition of the processes: "composting is the biological decomposition of wastes consisting of organic substances of plant or animal origin under controlled conditions to a state sufficiently stable for nuisance-free storage and utilization." Key components of the definition are "biological decomposition," "organic substances of plant or animal origin," "under controlled conditions," and "sufficiently stable."

A corollary of the specification "biological decomposition" is that only organic wastes of plant or animal origin can be broken down biologically. However, the restriction "only organic substances of plant or animal origin can be biologically decomposed" should be regarded as a generalization, because under rare circumstances certain microorganisms can attack material not of living origin. The requirement "under controlled circumstances" distinguishes composting from the ordinary decomposition that takes place in nature, in an open dump, or in a livestock feedlot. A problem with the specification "controlled conditions" is that of specifying the minimum control that would justify the designation "composting." Finally, the insertion of the requirement "stability" serves the twofold purpose of ensuring the production of an innocuous product and of providing a firm basis for making comparisons between compost systems.

## Biology

Most of the organisms active in the compost process are microbes. Groups of microbes active in the compost process are the bacteria and the fungi and some protozoa. The bacteria and the fungi are characterized by the successive appearance of their mesophyllic and thermophyllic forms. For purposes of convenience, the bacteria may be grouped into the bacteria proper and the filamentous bacteria. Of the filamentous bacteria, the actinomycetes are the most important in composting. The actinomycetes are most evident in the later stages of composting. Their appearance and proliferation is paralleled by the disappearance of cellulose and lignin. Nitrogen-fixing bacteria that may be present fix relatively insignificant amounts of atmospheric nitrogen, partly because of the slowness of the nitrogen fixation process, but mostly because of the abundance of available fixed nitrogen in wastes.

### *Vermicomposting*

If the compost process is allowed to continue beyond the minimum time required to attain stabilization, macrofauna begin to appear in the compost pile. The macrofauna utilize some of the microflora and decomposing waste as a substrate. Among the macrofauna, the earthworms (*Lumbricus terestris, L. rubellus, Eisenia foetida*) are perhaps the most conspicuous and certainly the most helpful. Unless they are deliberately introduced into a pile, these forms do not appear until the process is fairly well advanced. However, they can be introduced during the early stages of the compost process. The deliberate introduction and culture of such worms in the composting waste are the essential elements of a form of composting commonly known as "vermicomposting."

The activity of earthworms in composting material is said to have the following beneficial effects: (1) promotion and augmentation of particle size reduction; (2) removal of senescent (overage) bacterial colonies, thereby stimulating the proliferation and growth of new colonies; (3) enrichment of the composting material by way of nitrogenous excretions; (4) increase in microbial growth and activity through the minimization of bacteriostasis and mycostasis; (5) promotion of interparticulate penetration of atmospheric oxygen; (6) addition of mineral nutrients; (7) increase in the rate and extent of carbon and nutrient exchange because of influence on the interactions between microflora, protozoa, and nematodes; (8) amplification of pathogen control; and (9) production of worm castings. The castings have many beneficial characteristics of a good topsoil and are superior to compost as a soil amendment. The extent of the listed advantages varies from operation to operation.

A disadvantage of vermicomposting is the very careful control that must be exercised. In addition, the conditions required for vermiculture may not be universally available. Despite its seeming disadvantages, the possibility of using vermiculture as a part of composting (particularly at the household level) merits some consideration.

A diagram depicting the various stages and organisms involved in the composting process is presented in Figure 7.1. The figure also indicates the stages and organisms involved in the process of decomposition as it takes place in nature.

### *Inoculums*

With very few exceptions, the abundance and variety of microbes indigenous to wastes are sufficient to compost the wastes. An example of an exception would be a waste that for some reason had been exposed to pasteurization temperatures. A less likely exception could be a waste that is practically homogeneous in composition. Even with those exceptions, chances are that the required microbes would be present as a result of contamination during handling.

Because of the many different groups of microbes involved, but mostly because the compost process is a dynamic succession of microbial interactions with the substrate wastes, it would be a monumental task to isolate, identify, and determine the role of each group in the process. The three steps, isolation, identification, and determination of role, are all essential to the development of an inoculum. Moreover, carrying out the steps would tax the knowledge and competence of even a highly qualified microbiologist and certainly would be time consuming. Moreover, having isolated the organisms, their respective roles would be uncertain because of the difficulties involved in dealing with mixed cultures. Granted that if the microorganisms were identified and their role determined,

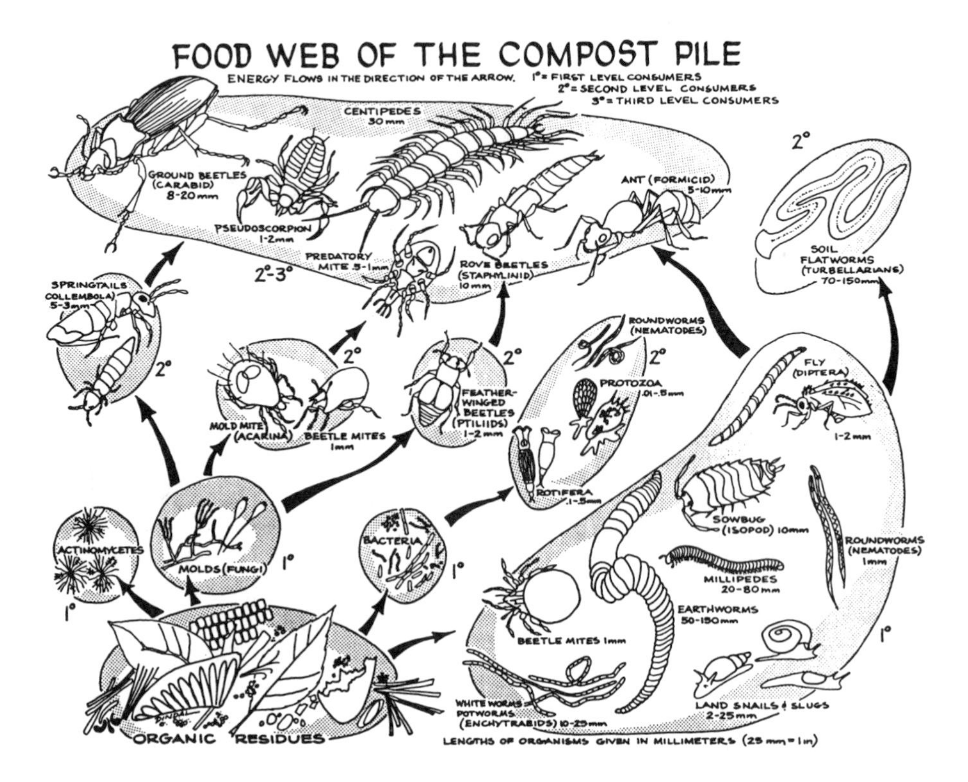


**Figure 7.1.** Stages of organisms involved in composting.

the ability of the inoculated bacteria to compete with indigenous bacteria and to thrive in an external situation would be markedly undermined by the repeated subculturing involved in maintaining a viable stock culture. Finally, no one has as yet unequivocally and scientifically demonstrated any efficacy of inoculums in composting. The upshot is that claims to the contrary should be viewed with a high degree of skepticism.

If for some reason an inoculum be desired, decomposed horse manure, finished compost, or a rich and loamy soil can serve the purpose. All three materials contain the required microbes.

As an outgrowth of wastewater treatment practice and industrial fermentation experience, mass inoculation of the incoming waste with finished compost often is advocated. Mass inoculation can take the form of recirculating a portion of the end product through the system. The efficacy of such a mass inoculation is debatable.

## Key Environmental and Operational Factors

Because microbes are the key active agents in composting, it follows that the factors that affect their proliferation and activity are those which determine the rate and extent of composting. Collectively, they are environmental in nature.

The substrate is one of the more important of the factors. Substrate-related factors are carbon-nitrogen ratio (C:N), particle size, oxygen ($O_2$) availability, aeration, moisture content, temperature, and pH. Of the preceding, chemical and physical nature of the substrate and aeration are especially important in process design. While strictly speaking, C:N, particle size, moisture content, and pH are all aspects of the nature of the substrate, in the present context, the term "nature of" is used to refer primarily to composition and availability of macro- and micronutrients in the substrate. A simplified diagram showing the major inputs and outputs of the composting process is given in Figure 7.2.

## Substrate

In composting, the substrate is the waste to be composted. Therefore, the discussion on nutrients which follows should be interpreted in that light. The waste should contain all necessary nutrients. Only rarely must or should chemical nutrients be added.

### *Nutrients: Types and Sources*

The macronutrients for microbes are carbon (C), nitrogen (N), phosphorous (P), and potassium (K). Among the micronutrients are cobalt (Co), manganese (Mn), magnesium (Mg), copper (Cu), and a number of other elements. Calcium (Ca) falls somewhere between the macro- and the micronutrients. However, the principal role of Ca probably is as a buffer, i.e., to resist change in pH level.

Even though nutrients may be present in sufficiently large concentrations in a substrate, they are unavailable to the microbes unless they are in a form that can be assimilated by the microbes and thereby become available to them. (The situation is analogous to that of cellulose in human nutrition, in that the carbon in cellulose is of no nutritional value to a person who might chance to ingest a cellulosic material, such as paper.)

An important point to remember is that availability is a function of the enzymatic make up of the individual microbe. Thus, certain groups of microbes have an enzymatic complex that permits them to attack, degrade, and utilize the organic matter as found in a freshly generated waste, whereas others can utilize only the decomposition products (intermediates) as a source of nutrient. The significance of this fact is that the decomposition and, hence, the composting of a waste is the result of the activities of a dynamic succession of different groups of microorganisms in which one group, so to speak, prepares the way for its successor group.

Another important aspect of nutrient availability in composting is that certain organic molecules are very resistant, i.e., refractory to microbial attack, even to microbes that possess the required enzymatic complex. The consequence is that such materials are broken down slowly, even with all other environmental conditions maintained at an optimum level. Common examples of such materials are lignin (wood) and chitin (feathers, shellfish, exoskeletons). Cellulose-C is

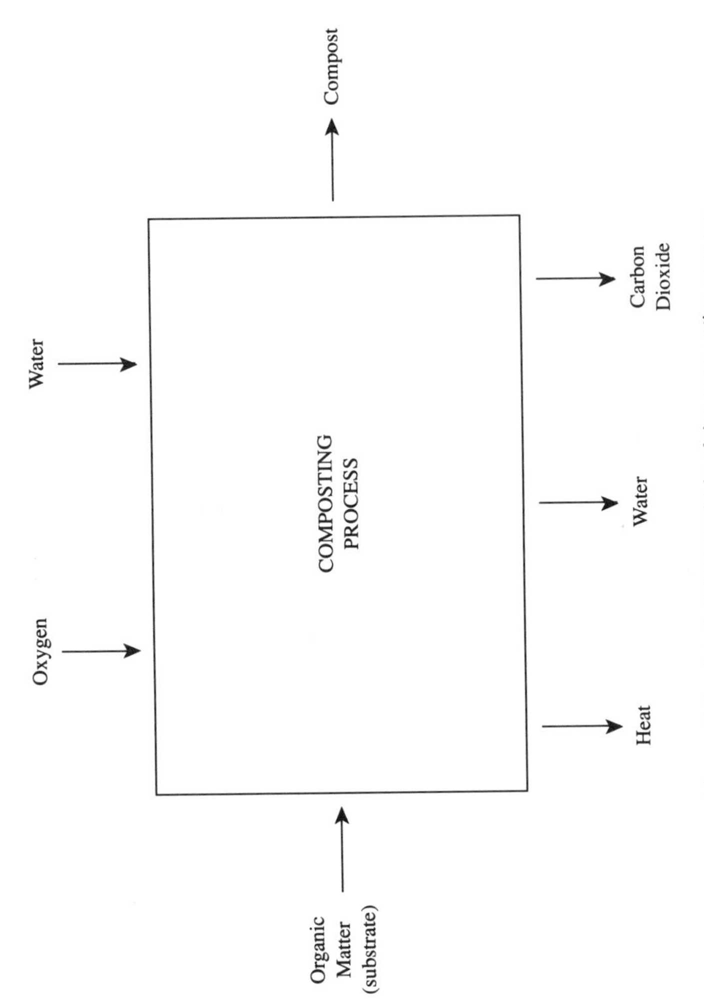


**Figure 7.2.** Input-output analysis of the composting process.

unavailable to the majority of the microbes, although it is readily available to certain fungi. Nitrogen is easily available when in the proteinaceous, peptide, or amino acid form, whereas the minute amounts present in chitin and lignin are difficultly available. Sugars and starches are readily decomposed, and fats are somewhat less so.

### *C:N*

With respect to the nutritional needs of the microbes active in composting, the C:N of the waste to be composted is the most important factor that requires attention. Experience shows that almost without exception all other nutrients are present in the typical organic waste in adequate amounts and ratios.

Requirements with respect to the C:N are functions of the relative differences in amounts of the two elements used by the microbes in metabolism to obtain energy and in the synthesis of new cellular material. A large percentage of the carbon is oxidized to carbon dioxide by the microbes in their metabolic activities. The remaining carbon is converted into cell wall or membrane, protoplasm, and storage products. The major consumption of nitrogen is in the synthesis of protoplasm. Consequently, much more carbon than nitrogen is required. The ratio is on the order of 20 to 25 parts of carbon to 1 of nitrogen. Therefore, the C:N of the substrate ideally should fall within the same range. Departures from the ratio of 20 to 25:1 lead to a slowing of decomposition and, hence, of composting. On the other hand, chances are good that nitrogen will be lost as ammonia-N if the C:N is lower than those levels. The reason for the loss is that nitrogen in excess of the microbial needs is converted by the organisms into ammonia. A combination of high pH level and elevated temperature very likely leads to volatilization of the ammonia.

If the C:N of a waste is too high, it can be lowered by adding a nitrogenous waste. Conversely, if the C:N is too low, a carbonaceous waste can be added. The nitrogen content and the C:N ratio of various wastes and residues are listed in Table 7.1. Additional information is presented in Chapters 6 and 8.

The nitrogen content of a waste can be determined by means of the standard Kjeldahl method. The determination of the carbon content is not as readily done because it is difficult to obtain a representative sample, the required analytical equipment is expensive, and an appreciable skill on the part of the analyst is demanded. Fortunately, an estimate of the carbon can be made that suffices for the purpose of composting. It is based on a formula developed in the 1960s by New Zealand researchers.[1] It is as follows:

$$\text{percent carbon} = \frac{100 - \text{percent ash}}{1.8}$$

Studies have shown that values obtained according to the formula approximate those from more accurate laboratory studies within 2 to 10%. In small-scale composting in which nitrogen and carbon analyses are not feasible, it can be

**Table 7.1 Nitrogen Content and C:N of Various Wastes and Residues**

| Waste | Nitrogen | C:N |
|---|---|---|
| Activated sludge | 5 | 6 |
| Animal tankage | | 4.1 |
| Blood | 10–14 | 3.0 |
| Cow manure | 1.7 | 18 |
| Digested sewage sludge | 2–4 | |
| Grass clippings | 3–6 | 12–15 |
| Horse manure | 2.3 | 25 |
| Mixed grasses | 214 | 19 |
| Night soil | 5.5–6.5 | 6–10 |
| Nonlegume vegetable wastes | 2.5–4 | 11–12 |
| Pig manure | 3.8 | |
| Potato tops | 1.5 | 25 |
| Poultry manure | 6.3 | 15 |
| Raw sewage sludge | 4–7 | 11 |
| Sawdust | 0.1 | 200-500 |
| Sheep manure | 3.8 | |
| Straw, wheat | 0.3–0.5 | 128–150 |
| Straw, oats | 1.1 | 48 |
| Urine | 15–18 | 0.8 |

assumed that the C:N will at least approach the proper level if the ratio of green (in *color*) fresh waste (or of food preparation wastes, or of fresh manure) to dry, nongreen waste is volumetrically about 1 to 4.

### *Particle Size*

The significance of particle size is in the amount of surface area of the waste particles exposed to microbial attack. It is a truism that the greater the ratio of surface area to mass (or volume), the more rapid is the rate of microbial attack. Therefore, theoretically, the smaller the particle, the more readily and rapidly it can be broken down. The term "theoretically" is used because size reducing certain fresh plant residues converts them into a slurry.

In practice, the minimum permissible particle size is that at which the porosity required for proper aeration in a composting mass can be attained and maintained. The maintenance depends upon the structural strength of the particles. Rigid, crush-resistant materials permit the retention of interstitial integrity even at very small particle sizes. On the other hand, fresh plant material tends to compress into a compact mass that has a very low degree of porosity.

In compost practice, the optimum particle size varies with the physical nature of the waste material. If the structure of the material is rigid and is not readily compacted, the particle size should be within the range of about 0.5 to 3 in. With fresh, green plant material, the minimum particle size of the greater part of the mass should not be less than 2 in., whereas the maximum particle size can be as much as 6 in., or even larger, depending upon the decomposability of the material.

The question often is posed as to the need for size reducing the organic fraction of waste. Generally, some size reduction is required, but whether or not a machine such as a hammermill would be needed is open to question. Perhaps, some form of tumbling in a drum could accomplish the relatively limited tearing, breaking, and maceration needed for wastes as highly organic and moist as are those generated in the preparation of food.

Some size reduction is indicated for woody or very fibrous residues to bring their particle length to 2 to 4 in. The longer particles would not interfere with the compost process per se, excepting perhaps to make handling more difficult. This assumes that the woody residues (branches and twigs) are no more than 0.5 in. in diameter, or if in the form of chips, no more than about 0.8 in. in width and 0.5 to 2.5 in. in thickness. Very little size reduction is required with fresh, green residues.

Unless the manure is intermingled with an abundance of bedding, it need not be size reduced. If size reduction is required, the requirement would be due to the bedding rather than to the manure.

## Environmental Factors

Because composting is a biological process, it is fundamentally affected by the collection of environmental factors that determine the course of action in all biological systems. The principal environmental factors of interest in composting are temperature, pH, aeration, moisture, and substrate (i.e., availability of essential nutrients). (Since the substrate was discussed in the preceding paragraphs, it will receive no further attention in this section.) These factors collectively determine rate and extent of decomposition. Obviously, the closer they collectively approach optimum levels, the more rapid will be the rate of composting. Despite the use of the term "collectively," the closest to which a given process can approach its maximum potential rate is determined by that factor which is farthest removed from being at an optimum level. The factor which places the uppermost limit is appropriately known the "limiting factor."

### *Temperature*

Much has been written about the relative merits of thermophilic versus mesophilic temperatures with respect to composting. Not unexpectedly, each temperature mode has its advantages and disadvantages; and probably the question of which is superior never will be settled to the satisfaction of the disputants. For practical compost operations, the dispute is largely academic because in the absence of strong countermeasures, both temperature levels will be encountered.

When the temperature rises beyond 150 to 160°F, the tendency is for spore formers to pass into the spore-forming stages. The transition is undesirable, because the spore-forming stage is a resting-stage, and therefore rate of decomposition is accordingly reduced. Moreover, microbes incapable of forming spores are strongly inhibited or even killed at those temperatures. Consequently, if it

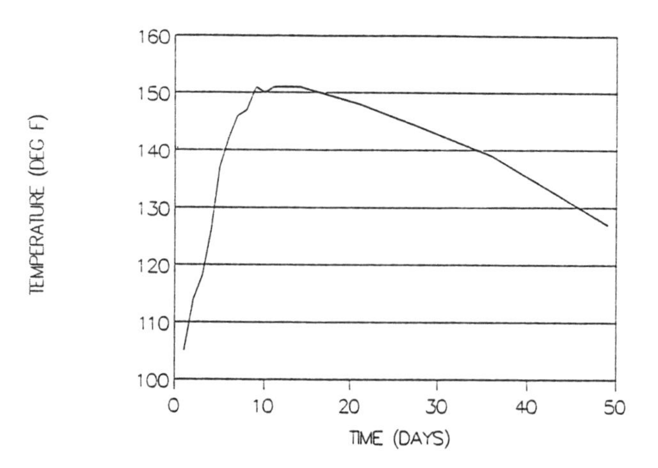


**Figure 7.3.** Typical temperature curve.

can be done, the top temperature should be kept at about 150°F. A typical temperature curve is presented in Figure 7.3.

The temperature distribution within a composting mass is affected by climatic conditions of its surroundings and by the method of aeration. In a windrow, the highest temperatures are developed in the center of the mass. The lowest temperature levels are reached at the corners of the pile, as shown in Figure 7.4. These temperature gradients promote a certain amount of natural air flow. The degree of air movement is a function of ambient conditions as well as porosity of the composting mass. Actual temperature profiles obtained in a windrow are presented in Figure 7.5. As shown in the figure, the temperature in the outside corners approaches that of the ambient.

### *Hydrogen Ion Level (pH)*

Whereas in anaerobic digestion, the critical pH level generally covers a fairly narrow range (e.g., 6.5 to 7.5), the range in composting is so broad that difficulties due to an excessively high or low pH level rarely are encountered. In composting, the pH level normally drops somewhat during the earliest stages of the process (i.e., down to 5.0) because of organic acid formation. The acids serve as substrates for succeeding microbial populations. Thereafter, the pH begins to rise, and may reach levels as high as 8.5.

Because it is unlikely that the pH will drop to inhibitory levels, there is no need to buffer the composting mass by adding lime (calcium hydroxide). Indeed, the addition of lime should be avoided because it can lead to a loss in ammonium

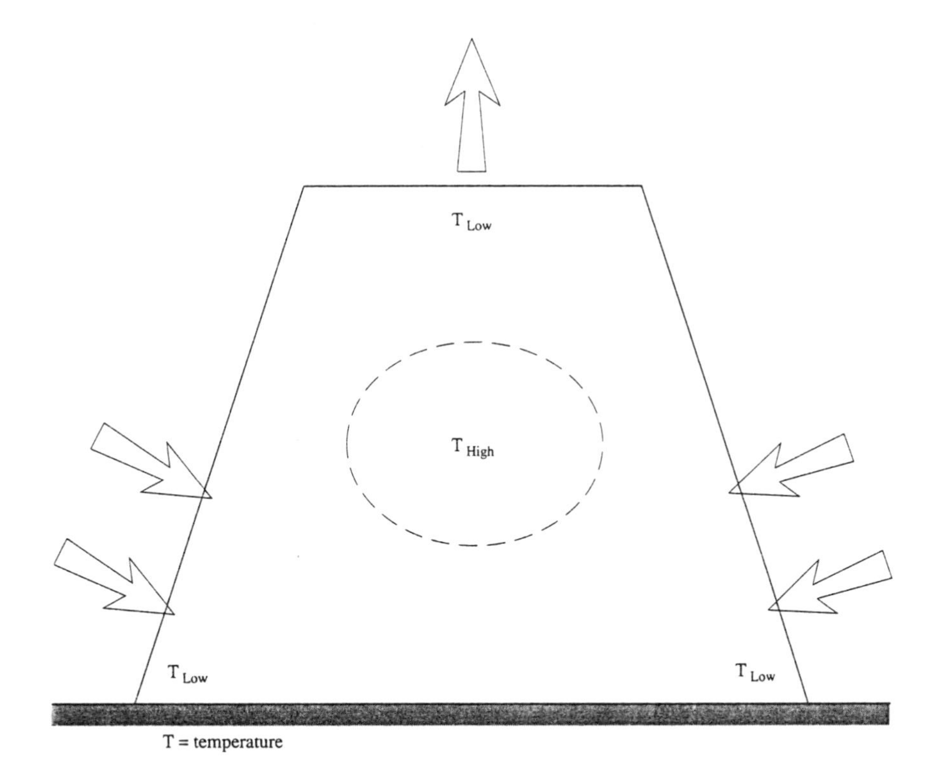


**Figure 7.4.** Natural air flow in a compost pile. T = temperature.

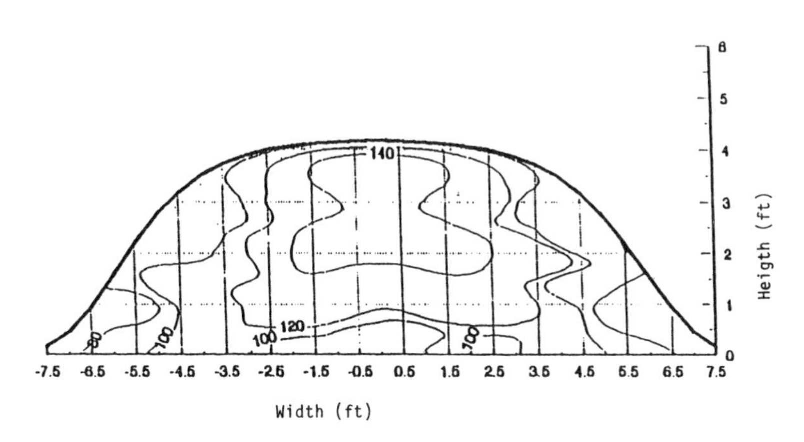


**Figure 7.5.** Temperature profiles in a windrow. Temperature in degrees Farenheit; wind speed: 20 to mph; ambient temperature: 55°F.

nitrogen in the later stages of composting that exceeds the normal loss. An exception could be in the composting of fruit wastes. With such wastes, the pH can drop to 4.5 There is some evidence that under such circumstances, the composting process can be accelerated.[2] At the relatively elevated temperatures and pH levels that occur as composting progresses, the ammonium ion is volatilized and the resulting ammonia gas is lost during the aeration of the composting mass. Although some loss of ammonia almost always occurs in aerobic composting, the loss is aggravated by the presence of lime. However, the lime does improve the physical condition of the composting wastes, perhaps partly by serving as a moisture absorbent.

### *Aeration*

At one time anaerobic composting was regarded as a serious alternative to aerobic composting. The major virtue attributed to the anaerobic mode of composting wastes was a minimization of the loss of nitrogen. Perhaps, with the increasing costs, financial and energetic, of nitrogen for fertilizing this advantage may be sufficiently significant to once again merit serious consideration. The second but much more dubious claim is that anaerobic composting is less costly than aerobic composting. The lower cost is more apparent than real, because the composting material must be thoroughly sealed off from the ambient environment and the time involved is much longer. The sealing-off is needed because of the offensive odors characteristic of anaerobically decomposing wastes. Other disadvantages include (1) slowness of decomposition, (2) absence of high temperatures, (3) the presence of undecomposed intermediates, and (4) the unprepossessing appearance of the product. The absence of high temperatures has a significance mostly in terms of public health in that the higher temperatures bring about a considerable destruction of disease-causing organisms. The presence of certain of the intermediates results in a compost product characterized by a generally foul odor.

Although it may somewhat increase the amount of nitrogen lost during composting, aerobic composting benefits from a relatively rapid rate of degradation, the attainment of elevated temperature levels, and the absence of putrefactive odors. It would be rash to claim that all odors produced in aerobic composting are not objectionable to all individuals, because chances are that the only universally unobjectionable odor is that of freshly turned loam.

Because the advantages of aerobic outweigh those of anaerobic composting, modern systems are almost without exception basically aerobic processes.

Several rates of aeration are mentioned in the literature. The number is due to the fact that rates required to keep the process aerobic are determined by the nature and structure of the wastes and by the method of aeration. For example, the combination of easily decomposed waste and a large and vigorous microbial population demands a larger amount of oxygen than does a combination of refractory materials and a sparse microbial population.

The determination of the amount of air required to ensure aerobiosis in a composting mass has been and is the goal of many researchers. The attainment of that goal is made exceedingly difficult by the fact that it can not be done by the methods of analysis developed over the years for wastewater.

One of the earlier investigations was carried on by Schulze.[3,4] Using a rotating drum as a reactor, he forced air through the drum at a given rate and measured the oxygen in the exiting air. Although Schulze's approach did not lead to a determination of the total oxygen demand of the material, it did give an indication of the rate of oxygen uptake. He found the respiratory quotient to be one that is

$$\frac{\text{carbon dioxide produced}}{\text{oxygen consumed}} = 1.0$$

In attempts to relate oxygen uptake to level of certain key environmental factors, he found that qualitatively the closer the environmental conditions approached an optimum level, the greater were the rate and amount of oxygen uptake. Thus, he reports that the oxygen uptake increased from 1 mg/g volatile matter at 86°F to 5 mg/g volatile matter at 145°F.[3,4,7,9] Conversely, oxygen uptake became less as environmental conditions worsened. The variability of results — a reflection of the variability of the composition of solid wastes — is illustrated by those obtained by Chrometzka[5] and Lossin.[6] Chrometzka reports oxygen requirements that range from 9 $mm^3$/g/hr for mature compost to 284 $mm^3$/g/hr with fresh compost. Lossin mentions demands that ranged from 900 mg/g/hr on day 1 of composting to 325 mg/g/hr on day 24. Regan and Jeris[7] report 1.0 mg oxygen per gram volatile solids per hour at a temperature of 86°F and a moisture content of 45% and 13.6 mg/g at a temperature of 113°F and a moisture content of 56%. A conclusion to be drawn from the results obtained by the preceding researchers is that, not surprisingly, oxygen uptake reflects intensity of microbial activity. Because of the lack of microbial activity, they concluded that it also is indicative of the degree of stability of the waste. If this were to be true, then the oxygen requirement would diminish after the composting mass has passed through the high temperature stage.

Theoretically, the amount of oxygen required is determined by the amount of carbon to be oxidized. However, it would be impossible to arrive at a precise determination of the oxygen requirement on the basis of the carbon content of the waste, since an unknown fraction of the carbon is converted into bacterial cellular matter and another unknown fraction is so refractory in nature that its carbon remains inaccessible to the microbes. The numerical value proposed by Schulze,[3] namely, 18,000 to 22,000 $ft^3$/ton of volatile matter per day, is useful with compost reactors equipped for metering air throughput.

The practical conclusion to be drawn from the preceding discussion is that experimentation should be done with the types of waste expected to be composted so as to determine rates of aeration. For windrow systems, the findings would

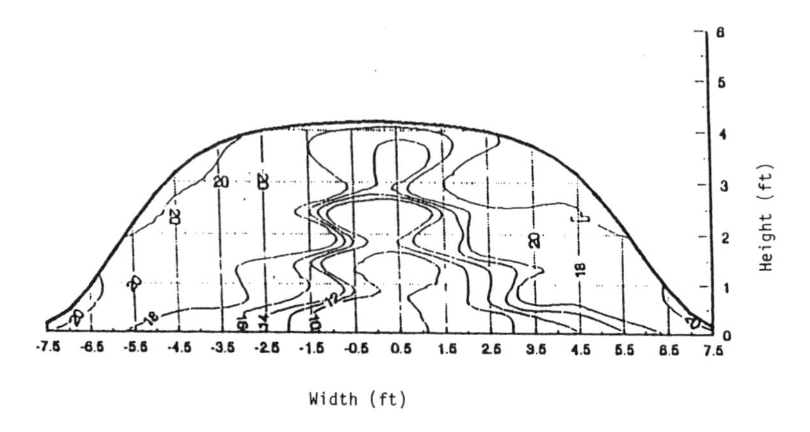


**Figure 7.6.** Concentration of oxygen in a windrow. Oxygen concentration = %; ambient temperature: 55°F; wind speed: 20 to 40 mph.

indicate the appropriate frequency of turning. For static piles and mechanized reactors, they would indicate rates of air flow.

Another practical conclusion is that in the absence of unusually strenuous efforts, it is impractical to maintain a completely aerobic state in a composting mass that is larger than about a ton. The aim should be to maximize aeration, but to do so within the constraint of financial feasibility. Because financial and, to some extent, technological constraints combine to impose a less than completely aerobic state, in general practice, oxygen availability could be a limiting factor.

The concentration of oxygen in a pile aerated mechanically is shown in Figure 7.6.

Conceivably, the compost process could be hastened considerably by enriching the input air stream with pure oxygen. In fact, such an approach has been seriously proposed. Although technically the concept seems to be attractive, it is highly doubtful that the returns would justify the sharply increased cost of doing so.

### *Moisture Content*

As stated earlier, permissible moisture content and oxygen availability are closely interrelated. The basis of the close relationship is in the method of carrying out the compost process. All methods involve the processing of the waste much in the state in which it is delivered to the compost site. In that state, it has a moisture content which is about the same as it had at the time it was generated. As such, the oxygen supply to the microbes involved is both the ambient air and the air trapped within the interstices (voids between the particles) of waste. In as much as the rate of diffusion of ambient air into the mass is inadequate, the interstitial air must be the major source of oxygen. Consequently, if the moisture content of the mass is so high as to displace most of the air from the interstices, anaerobic conditions (anaerobiosis) develops within the mass. There-

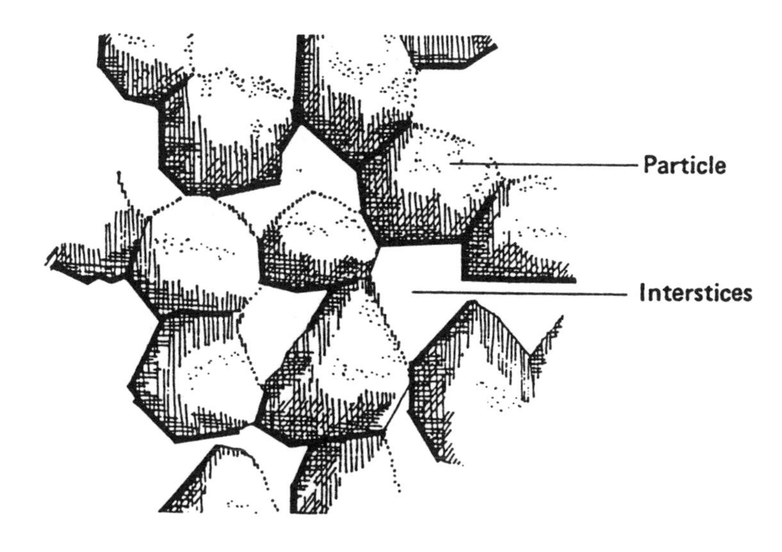


**Figure 7.7.** Schematic of interstices in composting material.

fore, the maximum permissible moisture content is the one above which the amount of air remaining within the interstices is not sufficient to assure an adequate supply of oxygen, i.e., oxygen becomes limiting. The term ''permissible'' implies a level at which no nuisance will develop and at which the process will proceed satisfactorily.

The maximum permissible moisture content is a function of the structural strength of the particles that make up the material to be composted. It refers to the degree of the resistance of individual particles to compression, i.e, the extent to which they maintain their original shapes upon being wettened or compressed. The compression refers only to that imposed upon particles by the weight of the mass above them. Obviously, the greater the structural strength, the higher is the permissible moisture content. Examples of such materials are woodchips, straw, hay (dried grass), rice hulls, and corn stover. Permissible moisture contents for mixtures in which such materials predominate are as high as 75 to 80%. If the particles are structurally weak, they are deformed (flattened) upon being subjected to compression, and the collective volume of the interstices is correspondingly reduced. The result is a lessening of the space available for air and water, and the permissible moisture content accordingly is lowered (see Figure 7.7). Paper is the principal example of such a material. Upon becoming wet, it collapses and forms mats. Mixtures in which paper is the major material have an upper permissible moisture content of only 55 to 60%. Finally, as a mass, the material to be composted may have little or no structural strength. For the sake of convenience, such wastes are referred to as being amorphous, i.e., the particles lacking a definite shape. Common examples are fruit wastes, cannery wastes, sludges, and manures devoid of bedding material. To compost those materials, it is necessary to add a ''bulking'' agent. A bulking agent is one that maintains its structural integrity when mixed with amorphous wastes. They also may have the capacity to absorb some moisture. Any material having a high degree of structural strength can serve as a bulking agent. In the absence of a bulking agent, an amorphous material can be subjected to a treatment such that it acquires a structural strength that is adequate for composting. For example, upon being dried, chicken manure takes on granular texture. Unless excessively wettened, the granular particles retain their integrity when mixed with fresh manure. In practice, some of the finished compost product is set aside for use as a bulking agent for the incoming waste.[9] An important point to keep in mind when using a dried or composted amorphous material as a bulking agent is that the combined moisture content of it and fresh waste should not exceed 60%.

The critical role of moisture content is not confined to windrow composting; it also applies to mechanized composting, including that in which the material is continuously agitated. As the moisture content of almost any mass is increased (i.e., up to the point at which it becomes a slurry), it takes on a tendency either to mat, to clump, or to form balls, or to do all three. By coincidence, the moisture content at which problems begin to be encountered are comparable to the upper permissible moisture contents for windrowed material.

In a discussion on moisture content, the lower levels at which moisture becomes limiting should receive attention. All microbial activity ceases when the moisture content is less than 8 to 12%. Consequently, moisture becomes increasingly limiting as it approaches that level. In practice, it is well to maintain the moisture content at a level above 40%.

## MONITORING THE PROCESS

### Course of the Process

To better understand the principles involved and to develop a suitable program for monitoring compost systems, it is important to describe the sequence of events that take place when all conditions are satisfactory. The description begins with the material in place, either in a reactor or in a windrow. Two obvious sequences are the rise and fall of temperature and the sequential change in appearance.

#### *Temperature Rise and Fall*

Normally, the temperature of the material begins to rise soon after the establishment of composting conditions, i.e., after the material has been windrowed or has been placed in a reactor unit. The initial rise in temperature is gradual (''lag period''). Immediately thereafter, if conditions are appropriate, the temperature rises almost exponentially with time until it begins ''to plateau'' at about 150 or 160°F. Depending upon the system used and the nature of the waste, the period of high temperature (''plateau'') persists for 1 to 3 weeks and then begins to decline gradually until ambient temperature is reached. If conditions are less than satisfactory, the high-temperature plateau may last much longer than 3 weeks, although the levels may be lower, e.g., at 130 to 140°F (see Figure 7.3).

The rise in temperature is due to two factors, namely, heat generated by the microbial population and the insulation against heat loss provided by the composting mass. The latter implies that at less than a critical mass, heat will be lost as rapidly as it is generated and the material will remain at ambient temperature. Under climatic conditions comparable to those encountered in coastal California (Mediterranean climate), the critical volume seems to be on the order of 1 cu yd. The critical volume is greater in colder climates — especially in regions where strong winds are common.

The heat generated by the microbes results from their respirational activities. Microbes are not completely efficient in converting and utilizing the chemical energy bound in the substrate. Energy not used becomes heat energy. Thus, temperature rise becomes an indicator of microbial activity, because the more active the microbial population, the greater is the amount of heat released. The exponential character of the rise in temperature is due to the breakdown of the easily decomposable components of the waste (e.g., sugars, starch, and simple proteins). It is during this period that the microbial populations increase exponentially in population size.

When the readily decomposable material has been composted, and only that which is more refractory remains, bacterial activity diminishes and, accordingly, the temperature drops. It may be assumed that by the time the temperature has descended to ambient or a few degrees above, the more biologically unstable components in the wastes have been stabilized and therefore the material is sufficiently composted for storage or for utilization, or for both.

### *Aesthetic Changes*

If the process is progressing satisfactorily, the composting mass loses the appearance it had as a raw waste and gradually assumes a darker hue. By the time the process is finished it has become a dark gray to brown in color. Change in odor is another eminently perceptible sequence. Within a few days, the odor of the raw waste is replaced by a motley collection of odors that, depending upon how well the process is advancing, range from a faint cooking odor to one redolent of putrefying flesh. The odors during this stage may be interlaced with the pungent smell of ammonia. If the C:N of the waste was low and the pH of the composting mass is above 7.5, the concentration of ammonia may mask other odors. Eventually, all objectionable and unobjectionable odors either disappear or are replaced by one that is suggestive of freshly turned loam. With respect to texture, the particle size tends to become smaller as a result of decomposition, abrasion, and maceration. Fibers tend to become brittle, and amorphous material becomes somewhat granular.

### *Molecular (Chemical) Changes*

A change not directly perceptible to the senses is the change in molecular structure. The change is manifested by a decline in concentration of organic matter and an increase in stability. (Organic matter often is referred to as "volatile matter" because combustion converts its carbon into carbon dioxide.) Because the compost process is a biological decomposition, oxidation of the carbon in organic solids to carbon dioxide is an important activity. Consequently, a portion of the organic matter is converted to carbon dioxide, i.e., is "destroyed." The controlled decomposition feature of composting makes it a degradative process in that complex substances are reduced to simpler forms. Complex molecules that are subject to biological decomposition (i.e., are biodegradable) are converted into simpler forms. Molecules that either are only partly or are completely unbiodegradable (i.e., refractory) tend to remain unchanged. The trend, then, is toward increased stability, in as much as a part of the decomposable mass is lost or reduced to simple forms and the refractory materials remain unchanged.

A trend has developed during the past decade to arbitrarily divide the compost process into two stages, namely, the "compost" ("active") stage and the "maturation" ("ripening," "curing") stage. The term "compost stage" applies to the period of rapid rise in temperature and may include the early plateau period. The term "maturation stage" includes the greater part of the plateau period and extends to and beyond the period of temperature decline. The division is strictly arbitrary in that composting takes place throughout the process, i.e., it is not discontinuous.

The division appeals to entrepreneurs for specific systems. With it, they can speak of "composting" as being done in terms of 2- or 3-day detention periods through the use of their particular systems. The accompanying 30- to 90-day maturation requirement is mentioned only in passing. Claims of 1- to 3-day composting are misleading. Consequently, upon considering claims by the purveyor of a particular reactor to the effect that only a 1- to 3-day detention period is required with the use of the reactor, one should inquire as to the length of the required maturation period. The reason for the query is that composting in the realistic sense of the term is not fully accomplished until it has progressed through the two stages. Most of these unreasonable claims have declined during the last few years.

## Parameters

A close review of the course of the compost process reveals four particular features that can serve as useful parameters for monitoring the performance of a compost system. They are (1) temperature rise and fall; (2) change in odor and appearance; (3) change in texture; and (4) destruction of volatile solids (i.e., organic matter).

The magnitude or intensity of the four features is much reduced if the wastes have a heavy concentration of inert material. Tertiary sludge is a good example.

The curve showing the course of the observed temperature rise and fall should resemble that of the curve in Figure 7.3. Upon being exposed to the appropriate operating and environmental conditions, failure of the temperature of the feedstock begin to rise rapidly within 1 to 3 or 4 days indicates that something is drastically amiss. A highly probable cause is too much or too little moisture. If malodors seem to be developing, then the problem is due to too much moisture. No odor is detected if the material is too dry. Another possible cause is an excessively high C:N. However, even with a high C:N, some increase in temperature should be detected. A pH at an inhibitory level could be a cause. Excessive moisture can be alleviated through the introduction of bulking material or by increasing the rate of aeration. Aeration removes moisture by way of evaporation. Obviously, a moisture shortage is eliminated through the addition of water. The addition of a highly nitrogenous waste (sewage sludge, and poultry, pig, or sheep manure) is the solution for a high C:N. A low pH can be raised through the addition of lime — keeping in mind the guidance recommended in the section on hydrogen ion level.

After the compost process has begun its course, a sharp deviation in any of the parameters mentioned in the preceding paragraphs indicates trouble. Thus, a sudden sharp dip in temperature during the period that normally would be time of exponential rise is a positive indication of the existence of some potentially serious problem. In a windrow, the dip generally is due to an excess of moisture. In a mechanical system, it might be malfunctioning of the aeration equipment. If excessive moisture is the cause, increased aeration (turning) is the best remedy for a windrow system. A more gradual but persistent decline during what should be the period of exponential rise or the "plateau" period is a sign either of inadequate aeration or of insufficient moisture.

The occurrence of objectionable odors invariably is a symptom of anaerobiosis caused by an excessively high moisture content or by inadequate aeration. In as much as the causative factor is anaerobiosis, the corrective measure is to increase the supply of oxygen. Consequently, the olfactory sense may be regarded as being an excellent device for monitoring adequacy of aeration. With a mechanized system, the olfactory sense can be supplemented by mechanical devices designed to monitor the oxygen concentration in the incoming and outgoing air streams. Basically, it should be possible to find some oxygen in the outgoing air stream. The presence of oxygen in the air discharged from a composting pile does not by itself imply adequate aeration. Aeration must be accompanied by proper air distribution throughout the pile.

The onset of a persistent decline in temperature, despite the continued presence of optimum conditions (i.e., no factor is limiting), indicates that the process is coming to an end and the composting mass is approaching stability. From past observations, it may be safely assumed that about the time the temperature begins to approach ambient temperature, the composting mass is sufficiently stable for storage and for use.

### *Determination of Degree of Stability*

Other than the final drop in temperature, no fully satisfactory test or analytical procedure is available for determining stability. The attainment of a dark color or an earthy odor is not an indication, because these characteristics may be acquired long before stability is reached. Attainment of a C:N lower than 20:1 is not indicative. With manures, the C:N of the raw waste itself may be lower than 20:1. Nor should dryness be confused with stability. It is true that if the moisture content is lower than 15 to 20%, microbial activity is minimal, and the product may seem to have the external attributes of stability. The actual situation, however, is that upon being wettened, the material has the degree of stability it had prior to dehydration. Although the fallacy of equating stability with dehydration may seem obvious, it nevertheless has been seriously offered as evidence by certain entrepreneurs.

Each of the considerable number of tests that have been proposed has some deficiency that seriously detracts from its utility. A deficiency found in every test is a lack of universality in terms of applicable values. Organic (volatile) solids concentration is an example. Ordinarily, it may be assumed that all substances having a comparable organic solids concentration are equally biologically stable. That assumption is not necessarily valid, because a substance containing organic solids of a refractory nature is more stable than one having an equal concentration of organic solids, but of compounds that are readily broken down.

Among the methods proposed for determining stability are the following six: final drop in temperature;[9] degree of selfheating capacity;[10] amount of decomposable and resistant organic matter in the material;[11] rise in the redox potential;[12] oxygen uptake;[13] growth of the fungus *Chaetomium gracilis*;[14] and the starch test.[15] The parameter, final drop in temperature, is based on the fact that the drop is due to the depletion of readily decomposed (unstable) material. This parameter has the advantage of being universal in its application. The course of the temperature (i.e., shape of the temperature curve) rise and fall remains qualitatively the same regardless of the nature of the material being composted.

Niese's analysis of self-heating capacity is a variation of the parameter, final drop in temperature.[10] In the conduct of his method, samples to be tested are inserted in Dewar flasks, which in turn are swathed in several layers of cotton wadding. Loss of heat from the flasks is further lessened by placing them in an incubator. Degree of stability is indicated by rise in temperature. The method has the universality of the parameter final drop in temperature. Its disadvantage is its slowness of completion. It may require several days to reach completion. Rolle and Orsanic designed their method to measure the amount of decomposable material in a representative sample.[11] The rationale for their test is that the difference between the concentration of decomposable material in the raw waste and that in the samples to be tested is indicative of the degree of stability of the latter. The basic principle involved in their test is that essentially, stability is a function of the fraction of oxidizable matter remaining in the composting mass. Hence, degree of stability reflects the size of the oxidizable fraction. The size of the fraction is determined by the amount of oxidizing reagent used in the analysis. Rolle and Orsanic's test consists of treating a sample with potassium dichromate solution in the presence of sulfuric acid. As a result of the treatment, a certain amount of dichromate added in excess is used up in the oxidation of organic matter. The oxidizing agent remaining at the end of the reaction is back-titrated with ferrous ammonium sulfate, and the amount of dichromate used up is determined. The calculation of the amount of decomposable organic matter is as follows:

$$\text{DOM} = (\text{ml})\ \text{N}\ (1 - \text{T/S})\ 1.34$$

in which DOM is decomposable organic matter in weight percent of dry matter and ml is milliliters of $K_2Cr_2O_7$ (potassium dichromate) solution. N is normality of potassium dichromate. T is ml of ferrous ammonium sulfate solution for back titration. S is ml of ferrous ammonium sulfate solution for blank test. Quantitatively, resistant organic matter is equal to the difference between the total weight lost in combustion and that degraded in the oxidation reaction.

The basis for Möller's test[12] is the reported incidence of human pathogens and parasites in incompletely composted material, and their supposed absence in the completed product. Theoretically, decomposable materials make possible an intensification of microbial activity and, hence, an accompanying increase in oxygen uptake, which in turn leads to a drop in the oxidation-reduction potential. Increase in mineralization of the composting material is paralleled by a rise in oxidation-reduction potential. According to Möller, stability has been reached if the oxidation-reduction potential of the core zone of the mass is $<50$ mV lower than that of its outer zone. He provides no method of testing material from a mechanized reactor. No zonation occurs in such a unit. An important obstacle to the use of the oxidation reduction potential is the lack of accuracy of the test and its vulnerability to a number of interfering factors.

Obrist[13] relies upon the effect had by degree of substrate stability on rate of growth and production of fruiting bodies of the fungus *C. gracilis*. According to him, growth of the fungus is dependent upon the chemical nature of the waste as a whole. The test consists in culturing the fungus upon a solid nutrient medium into which has been incorporated a pulverized sample of the compost and allowing the organism to incubate for 12 days at 100°F. At the end of the incubation period, the fruiting bodies are counted. According to Obrist, the more advanced the degree of stabilization, the fewer will be the fruiting bodies. Aside from the disadvantage of the overly long test period, is the serious one of requiring the services of an analyst skilled in mycological techniques.

The starch test seems to have gained a fair number of followers. According to Lossin,[14] its rationale is based upon the assumption that starch is always to be found in a waste, and that the concentration of starch declines with increase in degree of stabilization. Lossin states that three types of carbohydrates occur in wastes, namely, sugars, starch, and cellulose. Not unexpectedly, during the course of composting, sugars are the first to disappear, followed by starch, and finally by cellulose. Lossin reasons that because starch is relatively easy to break down, and all wastes contain starch, no starch should remain if a composted material is to be considered stable. The analysis is based upon the formation of a starch-iodine complex in an acidic extract of the compost material. Care must be exercised in conducting the test because it is easy to arrive at false results. The test again suffers from the inability to set up universally applicable values.

Recently, a study was made of the comparative reliability of some of the tests.[15] One of the primary objectives of the study was to find methods by means of which the degree of decomposition of a compost could be quickly and reliably determined. The "transferability" and universality of the methods were to be determined and compared. Findings made in the study showed that of the methods investigated, self-heating, oxygen consumption (4 days), and ratio of oxygen consumption to chemical oxygen demand provided the most reliable results.

## TECHNOLOGY

The technology of composting involves the preparation of the raw material ("preprocessing"), the compost process itself, and the grading and upgrading of the final product ("postprocessing"). The preparation steps are size reduction and sorting. In as much as size reduction and sorting are discussed in Chapter 4 (Processing), this chapter is concerned solely with the compost and product preparation steps.

### Principles

The discussion of the various principles upon which rests the technology of composting is begun by emphasizing that if a proposed compost facility is to have a capacity larger than a few tons per day, it should be designed by professionals who are thoroughly conversant not only with solid waste management, but also with the theoretical and practical aspects of the composting. Many unsuccessful ventures are attributable to a failure to sufficiently involve competent professionals.

#### *Purpose of Equipment*

The general purpose of all equipment in composting (as with any biological system) is to provide the microbial agents with an optimum environment and to do so within the constraints of financial feasibility. The general experience in composting is that oxygen availability is the environmental factor of most concern. Consequently, the emphasis in equipment design for composting has been on the development of effective aeration systems.

### *Aeration*

As stated in the section "Moisture Content," the source of oxygen for the microorganisms is the layer of air that more or less envelopes each particle. Oxygen removed from the surrounding air is replaced by carbon dioxide released by the microbial cells. Eventually, the supply of oxygen in the envelope of air is depleted, and unless it is replaced by an envelope of untapped air, anaerobic conditions soon prevail. Therefore, the ultimate aim in the design of aeration equipment is the renewal of the gaseous environment at a rate such that a supply of oxygen is always available. The renewal is accomplished either by physically moving the particles into new position and thereby exposing them to fresh supplies of air, or by displacing the gaseous envelope and allowing the particles to remain stationary.

*Physical* movement of the particles (*agitation*) is accomplished in two ways, namely, by tumbling and by stirring. In tumbling, the particles are lifted and then allowed to fall or drop. Renewal of the gaseous envelope is accomplished during the falling movement. Obviously, the dropping should be accomplished such that the composting material is not compacted. Tumbling may be accomplished by "turning" the material, as is explained in the section on windrow composting. In mechanical compost systems (see section, "Mechanical Systems"), it is accomplished by one or more means. In some systems, tumbling is brought about simply by dropping the composting mass from one floor to another or from one conveyor belt to a lower one. In others, tumbling is accomplished by introducing the material into a rotating cylinder equipped with interior vanes. In systems involving stirring, movement is primarily sideways, and tumbling is practically nonexistent.

In a system in which the particles remain stationary, the air layer is constantly diluted or replaced by air forced through the composting mass; hence, the term "forced aeration" is applied to such a system (the adjective "static" is frequently used in place of "forced aeration"). The effectiveness of this approach depends upon the even distribution and unobstructed movement of air to all parts of the composting mass.

## Guidelines to Selection of Equipment and Systems

Certain general guidelines exist which, if followed, will minimize the number of undertaking of projects that are inevitably doomed to failure or prove to be needlessly burdensome. The guidelines offered herein are the outcome of four decades of first-hand experience and observation.

The first, and extremely important, guideline is more a principle than a guideline. It is that to be successful, a compost system need not be complex. It is based upon the tenet that complexity and efficiency of process are not mutually dependent. In fact, excessive complexity in composting all too often leads to gross inefficiency.

Although especially applicable to developing nations, the second guideline is not without validity in industrialized nations. The essence of the principle is that the selected system be readily adaptable to the economic and manpower conditions of the locale in which it is to be used. Good sense dictates that a thoroughly automated compost system not be selected for a nonindustrialized situation that also is complicated by a high unemployment rate and a dearth of available or accessible foreign exchange. This is especially true in such a situation, because chances are that a simplified, labor-intensive system might better serve the purpose. Under the circumstances that generally prevail in a nonindustrialized situation, a sophisticated compost installation soon becomes unworkable due to the lack of skilled labor, adequate maintenance, and replacement parts. Incidental nuisances that might accompany the use of a low-technology system can be screened from the public by a suitable yet simple shelter.

The third guideline is more in the nature of a series of "caveats" rather than of a guideline and is especially applicable to enclosed systems. The precautions are suitable equally for the average person in private life and for public undertakings. The first precaution is to the effect that exaggerated claims made by an entrepreneur should be regarded with substantial skepticism.

Regarding compost systems, claims tend to become unrealistic when they deal with acceleration of the process, magnification of efficiency, and production of a superior product. For example, claims for 1-day or even 6-day composting clearly are exaggerated claims. Even with a highly putrescible waste and with fairly effective aeration, a minimum of 12 to 14 days of total composting time is needed. A careful examination of the claims will show that they call for an abbreviated residence time in the "reactor" unit followed by an extensive retention as a windrow. The highly unequal division of retention times fits in with the previously mentioned trend to split the compost process into a "compost stage" and a "maturation stage." Upon reviewing the claims for rapidity, one cannot fail to note a correlation between high cost of reactor and brevity of the claimed compost time. It becomes apparent that the claimed increase in rapidity is more a function of not discouraging the prospective purchaser, rather than of any intrinsic superiority of the reactor. The important point to keep in mind is the total of the times named for the "compost" and the maturation stages. As it becomes apparent in the sections which follow, in most situations the time gained, if any, by sophistication of equipment is not great enough to offset the substantial additional expenditure that is required.

The best way to evaluate a given system is to have it observed while it is in action by an individual who is thoroughly versed in waste processing and composting, as well as to inspect the product directly as soon as it is produced, i.e., the same day. Samples kept on hand for later inspection are not truly representative because they benefit from prolonged storage and from a certain amount of drying. Dryness conceals instability. In order to be able to detect hidden weaknesses and potential pitfalls, the observer should be a professional who is completely conversant with composting through firsthand experience. The observation period should be a minimum of eight consecutive hours.

Simplicity or complexity of process has little effect on the quality of the product, provided the process is properly conducted. However, the product can be upgraded by further processing.

At this time, attention is called to the remote possibility that a reactor could be designed such that a product could be produced within the retention times claimed even by the more optimistic entrepreneurs. Unfortunately, the capital and operating costs of the resulting setup would be extremely high. An example of such an approach would be to make an ultrafine slurry of the waste and then subject the slurry to the conventional activated sludge process used in wastewater treatment. Versions of this approach are being applied to the stabilization of wastewater sludges.

In summary, it should always be kept in mind that composting, being a biological process, is constrained by the limitations of a biological system. As was stated earlier, a process proceeds at a rate and to an extent equal only to those permitted by the genetic traits of the microorganisms. No amount of sophistication of equipment can bring about a further increase.

## Types of Systems

The classification of systems is based upon degree of mechanization of the compost method and is expressed as a loose division into two broad groups, namely, windrow (open) and mechanical or enclosed, with a considerable overlap between them. The names are self-explanatory, although it should be pointed out that windrow systems can be mechanized to a considerable extent and may even be partially enclosed. Windrow systems are further subdivided under the headings "static" and "turned."

### *Static System*

With the static system approach, air either is forced up through the composting mass or is pulled down and through it, hence, the alternative designation "forced aeration." In both cases the composting material remains undisturbed.

Despite the fact that the forced air system of aeration had been proposed and tried as early as in the late 1950s,[16] it was not until the 1970s that it began to receive a considerable attention. Although Senn[8] successfully applied it in the composting of dairy cattle manure, the primary reason for the resurgence of interest was the apparent utility in the application of the method to the composting of sewage sludge. The system as applied to sewage sludge is the one known as the Beltsville method of composting — so-called after the name of the place of its origin.[17] The system essentially involves an initial period of pulling air into and through the pile followed by a period of forcing it upward through the pile. In the pulling or "suction" stage, the exiting air either is not discharged directly

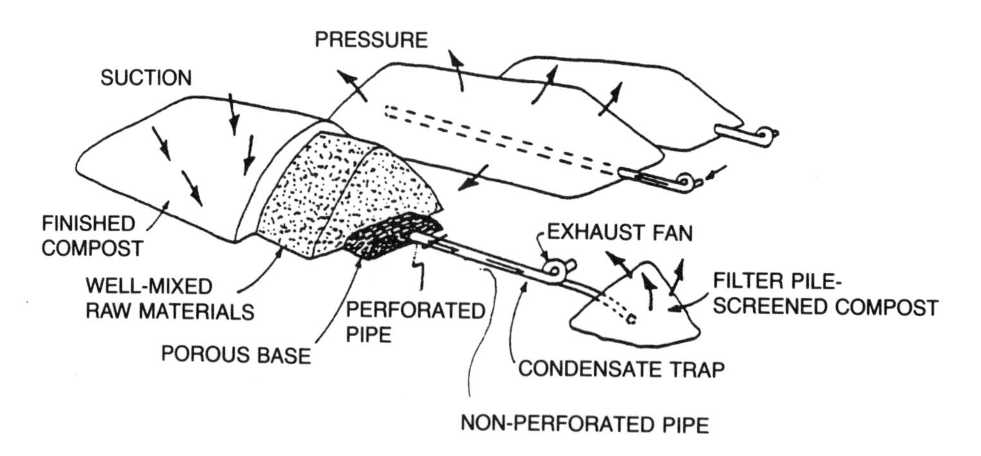


**Figure 7.8.** Aerated static pile composting — individual piles.

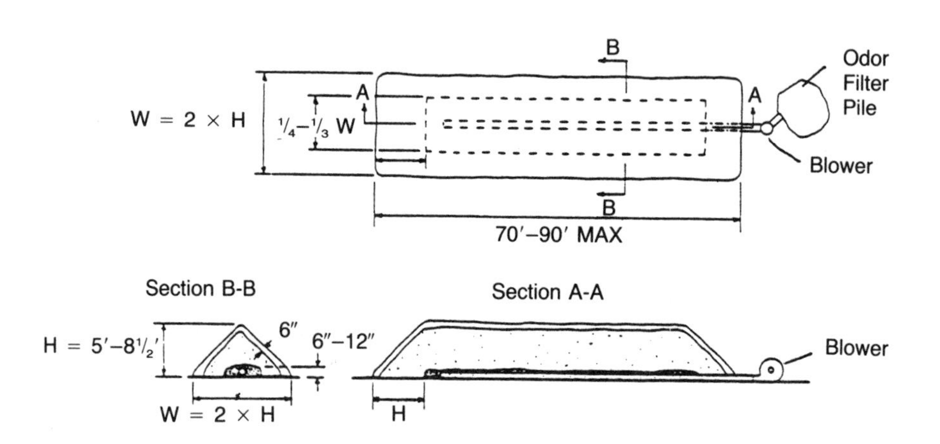


**Figure 7.9.** Approximate dimensions for an aerated static pile.

into the environment or is forced through a pile of finished compost or other "stable" organic matter. The rationale for the latter procedure is to deodorize the effluent air stream. Apparently, finished compost can serve as an odor filter. The basic arrangement of an aerated pile is shown in Figures 7.8 and 7.9.

The system includes the following main steps: if a bulking agent is required (e.g., for sewage sludge, manure devoid of bedding material), it is mixed with the waste to be composted. The second step is the construction of the windrow (i.e., elongated pile). The third step is the composting process. The fourth step is the screening of the composted mixture to remove reusable bulking agent. Curing and storage, respectively, constitute the fifth and sixth steps.

The construction of the windrow proceeds as follows. A loop of perforated pipe, 4 to 6 in. (diameter), is placed on the compost pad. The loop is oriented

**Table 7.2 Typical Specifications for Aeration System**

| | |
|---|---|
| Piping grid | |
| Perforated length[a] | 50 to 75 ft |
| Diameter | 4 to 6 in. |
| Air handling | |
| Blower size | 1/3 to 5 HP |
| Flowrate | 25 to 100 cfm/dry ton |

[a] For even spacing and uniform hole size.

longitudinally and is centered under what will be the ridge of the windrow. Short circuiting of air is avoided by ending the pipes about 5 to 9 ft from the edges of the windrow. The perforated pipe is connected to a blower by means of a length of nonperforated pipe. After the piping is in place, it is covered with a layer of bulking agent or finished compost that extends over the area to be covered by the windrow of material to be composted. This foundational layer is provided to facilitate the movement and uniform distribution of air during composting. It also absorbs excess moisture and thereby minimizes seepage from the pile. The material to be composted is then stacked upon the piping and bed of bulking material to form a windrow of the configuration shown in Figure 7.8, i.e., roughly triangular in cross-section. The finished pile should be 70 to 90 ft long, is about 10 to 18 ft wide, and is about 5 to 9 ft high.

Finally, the entire pile of composting material may be covered with a layer of matured (finished) compost that is 6 in. thick if the covering compost is screened, and 8 in. if unscreened. The covering serves to absorb objectionable odors from the composting mass and to ensure the occurrence of high-temperature levels throughout the composting material. The arrangement accomplishes a more complete pathogen-kill than would otherwise take place. Experience indicates that a continuous forcing of air through the pile is not needed for aerobic conditions to persist. For example, in a study which involved an 80-ton pile, a timing sequence proved to be adequate that involved forcing of air into the pile at 9 CFM for 5 to 10 min at 15-min intervals. This particular rate was based upon a need of about 4 L/sec/ton of dry sludge solids. It should be noted that these numbers are only indicative. For a given operation, the required rate of air input should be determined experimentally, because it depends upon a number of variable factors.

General specifications for the aeration system are given in Table 7.2.

Because porosity is a key factor in forced aeration, it is important that the moisture content be such that the interstices or voids be devoid of water. A safe level of moisture content is one within the range of 40 to 55%.

The effluent air is passed through a small, cone-shaped pile, preferably of matured compost. At Beltsville, these piles usually were about 4 ft high and about 8 ft in diameter at the base. The moisture content of the material in the piles should be less than 50%.

In the Beltsville process, the sludge to be composted (approximately 22% solids) is mixed with woodchips (bulking agent) at a ratio of 1 volume of sludge to 2 volumes of woodchips. The compost process requires from 2 to 3 weeks. When the process has reached completion, the pile is torn down and the material is screened. If the bulking material is to be recycled, the screen-opening size should be such that the bulking particles are retained on the screen and compost passes through. Because wet material is difficult to screen, screening should not be performed during rainy days.

If large amounts of material are to be composted, the so-called "extended aerated" pile can be used. The program followed with such an arrangement is as follows. On day 1, a pile is constructed in the manner described in the preceding paragraphs, excepting that only one side and the two ends of the pile are blanketed with the matured compost layer. However, the exposed side is lightly covered with matured compost to prevent the escape of objectionable odors. On day 2, a second loop of piping and bedding are laid directly adjacent to the exposed side of the pile erected on day 1, and the pile is erected in the same manner as was pile 1. This procedure is repeated for 28 days. The first pile is removed after 21 days, the second pile on the day after, and so on. An important advantage of the extended approach is a substantial reduction in spatial requirements.

The land area requirement for single-pile systems is about 1 ha per 7 to 11 tons (dry weight) of sludge processed. The 7 ton/ha estimate allows for land area to accommodate runoff collection, administration, and general storage.

*Economics and Limitations.* Of the various types of compost schemes, the static pile method is perhaps the least expensive, especially when the volume of the raw material is greater than can feasibly be handled by manual labor in a region in which unemployment is chronic, and certainly so when labor is both scarce and expensive. The reasons for the low costs are readily apparent, namely, limited amount of materials handling required and the relatively inexpensive equipment required.

It is difficult to arrive at a generally applicable capital cost for static pile composting, because it is site specific. With respect to material and operational costs, the cost for composting a sludge and woodchip mixture is about $50/ton (1991 dollars), of which about $10 is for woodchips. The cost of woodchips, however, can be much higher in some regions.

The method is not suitable for all types of raw materials and under all conditions. For instance, it works best and perhaps only with a material that is relatively uniform in particle size and in which the particle size does not exceed 1.5 to 2 in. in any dimension. Granular materials are best. Overly large particles and a wide distribution of particle sizes lead to an uneven distribution and movement of air through the pile, thus leading to short-circuiting and the development of anaerobic pockets of decomposing material.

### *Turned Windrow System*

The turned windrow method is the one traditionally and conventionally associated with composting. The term "turned" or "turning" applies to the method used for aeration. In essence, it consists in tearing down a pile and reconstructing it. The details and variations in methods of turning are many. In addition to promoting aeration, turning ensures uniformity of decomposition by exposing at one time or another all of the composting material to the particularly active interior zone of a pile. To some extent, it may also serve to further reduce the particle size of the material. A dubious advantage is the loss of water that is accelerated by the turning process. The loss is a definite advantage if the moisture content is overly high. Conversely, it is a disadvantage if the moisture level is too low. An excellent time to make a necessary addition of water is during the process of turning.

*Construction of Piles.* The pile should be in the form of a windrow and generally, roughly conical in cross-section. However, under special circumstances the cross-sectional shape should be adjusted to fit those circumstances. For instance, during dry, windy periods, a loaf shape tending toward a flattened top would be appropriate because the ratio of exposed surface area to volume is lower with such a configuration. Moreover, the volume of the overall hot zone is greater than with a triangular or conical cross-section. On the other hand, the flattened top becomes a distinct disadvantage during wet weather because water is absorbed rather than shed. In operations in which the turning is done mechanically, the resultant pile configuration is the one imparted by the machine.

Ideally, the windrow should be about 6 ft high. In situations in which it is practical to do the turning manually, the height should be roughly that of the average laborer. At most, it should be no higher than that easily reached with the normal pitch of the instrument used in turning. Another determinant of maximum height is the tendency of stacked material to compact. The height for mechanical turning is a function of the design of the turning equipment. Generally, it is about 7 ft.

The breadth of the pile is determined by convenience and expediency. The reason for the latitude is that diffusion of oxygen into a pile is a minor element in meeting the oxygen demand of the composting mass. With manual turning, a width of about 8 ft seems to be suitable. With mechanical turning, the width depends upon the type of machine. Usually, it is from 10 to 13 ft.

The length of the windrow is indeterminate. For example, the length of a 180-ton windrow (conical shape) of material at a height of 6 ft and width of 8 ft would be about 150 ft. A quasicontinuous system can be established by suc-

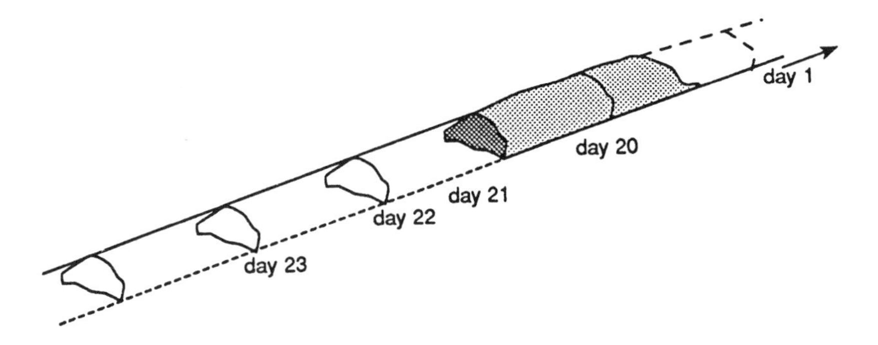


**Figure 7.10.** Construction of piles.

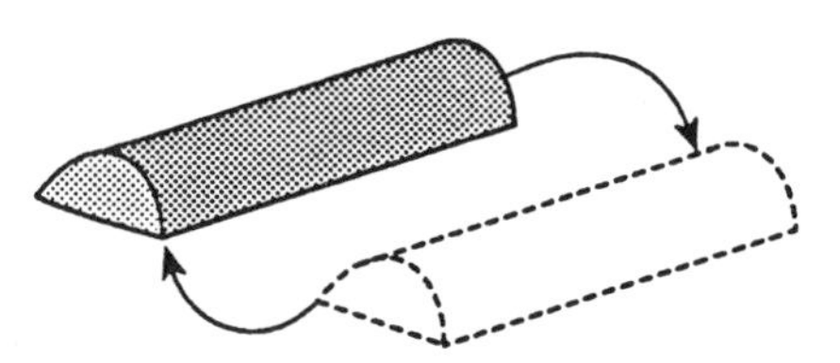

**Figure 7.11.** Logistics of turning windrows.

cessively adding each day of input of raw waste to one end of the windrow, as shown in Figure 7.10. In essence, it consists in adding fresh material to one end of the windrow and removing material from the other as it reaches stability — i.e., is composted.

*Arrangement of the Windrows.* The arrangement of the windrows at the compost facility depends upon accessibility by the equipment. Whatever the arrangement, the windrows should be positioned such that the course of each day of input can be followed until it is completely composted.

An important requirement is the space needed to accomplish the turning of a single day of input, whether it be done manually or mechanically. With manual turning, the total areal requirement is at least 2 times and more likely 2.5 times that of the original windrow. This follows from the logistics of turning, as is indicated by the diagram in Figure 7.11. When the second turning takes place, the windrow is returned to its original position. This double space requirement for each day of increment continues until the material reaches stability.

The areal requirement in mechanical turning varies with type of machine. Some machines accomplish turning such that as the original windrow is torn down, it is reconstructed directly behind the machine. This is done either by designing the machine such that it straddles the windrow, or that it has a mechanism for tearing down a windrow and passing the material over the cab to the

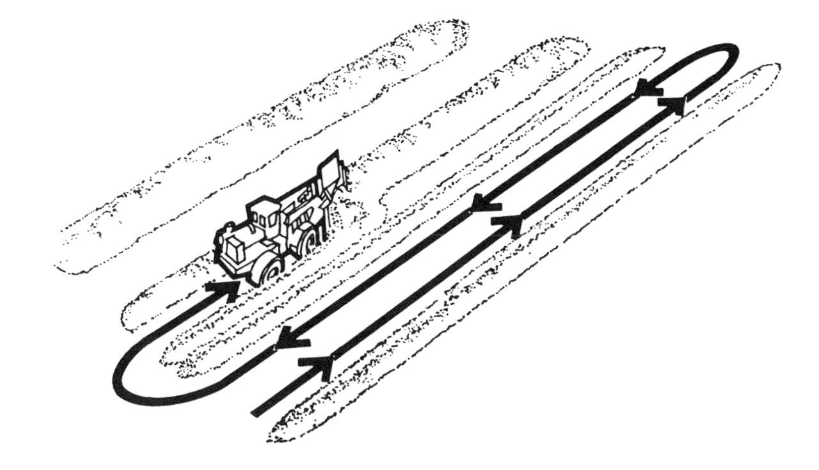

**Figure 7.12.** Arrangement for mechanical turning.

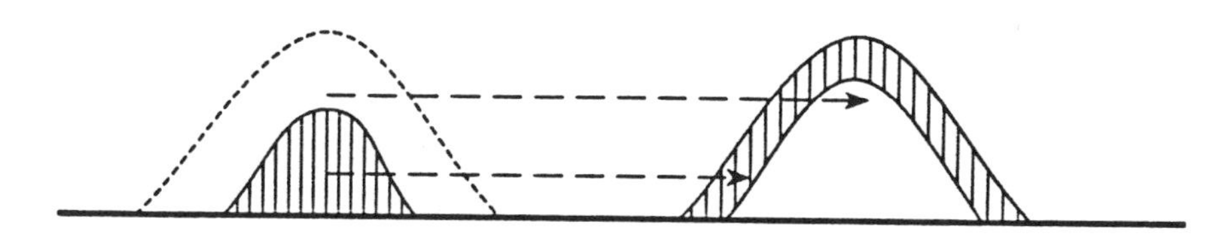

**Figure 7.13.** Method of rebuilding a pile.

rear of the machine as it moves forward. The areal requirement with such a machine is little more than that of the original windrow, including only enough added space to permit the maneuvering of the machine.

Other types of machines rebuild the windrow adjacent to its original position. The areal requirement therefore is comparable to that described for manual turning. The arrangement for such a machine is shown in Figure 7.12.

*Methods of Turning.* The tool most convenient for use in manual turning is a four- or five-tined pitchfork. Certain things should be remembered when turning a windrow manually. Ideally, in rebuilding the pile, material from the outside layers of the original windrow should become the interior of the rebuilt windrow, as is shown in Figure 7.13. Because in practice it is not always convenient to turn the windrow in such a manner, during the compost cycle every particle of material should be at one time or another in the interior of the pile. If this ideal is unattainable — as is the case with mechanical turning — the deficiency can be compensated by increasing the frequency of turning, i.e., by turning two or even three times per week. Finally, when constructing the original windrow, and when rebuilding it, care should be taken not to compact the material.

*Frequency of Turning.* Ultimately, the frequency of turning is indicated by the ratio of oxygen availability to oxygen demand. In a practical situation, it is a compromise between need and economic and, to some extent, technical feasibility of supplying that need. Nature of the material, especially its structural strength and moisture content, is important in determining the frequency of turning. Factors in addition to effectiveness of the turning procedure are pathogen kill and uniformity of decomposition. A variable factor is the rapidity of decomposition desired by the operator. High-rate composting demands very frequent turning because, to a certain extent, rate of composting is directly proportional to frequency of turning. The drier the material and the firmer the structure of the particles, the less frequent will be the indicated turning.

Judging from the authors' experience,[9,18] when straw, rice hulls, dry grass, dry leaves, woodchips, or sawdust constitute the bulking material and the moisture content is about 60% or less, turning on the third day after constructing the original pile and every other day thereafter for a total of about four turnings is sufficient to accomplish "high rate" composting. After the fourth turning, the frequency need be only once each 4 or 5 days. The same program is applicable if paper is the bulking material (e.g., U.S. municipal refuse), provided that the moisture content does not exceed 50%.

If the composting mass gives off vile odors, it is anaerobic and, hence, requires further turning. Usually, the onset of anaerobiosis is triggered or is hastened by the presence of excessive moisture. Because it fosters evaporation, increasing the frequency of turning to at least once each day soon results in the disappearance of the odors.

*Equipment (Turning).* A rototiller of the type pictured in Figure 7.14 is quite satisfactory for small operations. Turning by means of a rototiller is accomplished by tearing down the pile and spreading the composting material to form a 12- to 24-in. layer. The rototiller is then passed back and forth through the layer. The routine should be such that the operator does not tread upon the agitated material and thereby compact it. After the material is agitated, it is reconstituted into another pile. Because of the small capacity of the machine, and the nature of its manipulation, the rototiller is confined to small-scale operations.

Several types of machines specifically designed to turn refuse are on the market. The Brown Bear is the exception, in that it was originally designed to "backfill" ditches. The machines among themselves differ in degree of effectiveness and durability. The list that follows is a sampling of machines manufactured and sold in the U.S. They are the Scarab, the Cobey Composter, the Brown Bear, the Scat, the Wildcat, and the King of the Windrow. These machines can process from a few to 3000 tons/hr of fresh compost. An idea of the general design of the machines may be gained from the diagrammatic sketch in Figure 7.15. Costs (1991) of the self-powered machines range from about $100,000 to

**Figure 7.14.** Photograph of rototiller.

about $185,000. A partial listing of windrow turning equipment self-powered and PTO driven is presented in Tables 7.3 and 7.4, respectively.

*Site Preparation.* The piles should be placed on a hard surface, preferably a paved surface during the active stages of composting. Reasons for the paving are (1) to facilitate, or even to make possible materials handling; (2) to permit control of any leachate that may be formed; and (3) to prevent fly larvae from escaping the area. In summary, preservation of sanitation and materials handling are the two decisive factors. With small-scale operations (less than about 10 tons/day), the paving may consist simply of compacted clay as a base with packed gravel or crushed stone on the surface. Where crushed stone and gravel are not available, the soil should at least be firmly packed. Of course, in the latter case, a problem arises during the rainy season. Paving is especially essential if mechanical turners are utilized. The machines are fairly heavy and, accordingly, can operate properly only on a firm footing. Paving materials in addition to gravel and crushed stone are asphalt and concrete.

Provision should be made for collecting the leachate that might be generated. Such leachate has an extremely objectionable odor and, unless controlled, can lead to the development of problems. In desert regions, the windrows should be protected from the wind so as to reduce moisture loss through evaporation. In regions of moderate to heavy rainfall, the windrows should be sheltered from

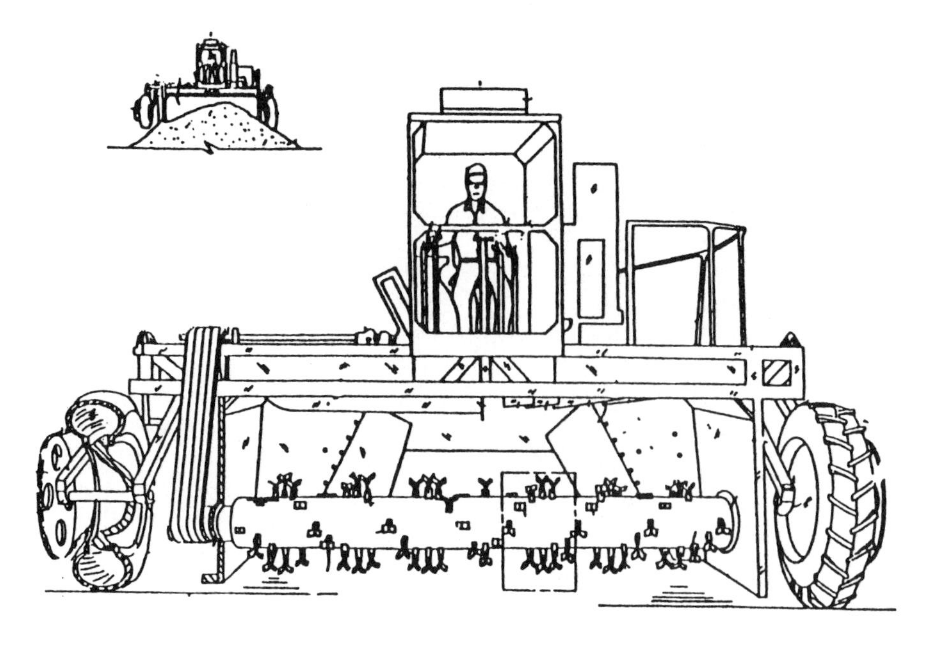

**Figure 7.15.** General design of mechanical turners.

**Table 7.3 Partial List and Cost of Windrow Turning Equipment (Self-powered)**

| Manufacturer | Power (HP) | Capacity (TPH) | Approximate Cost (U.S. $1991) |
|---|---|---|---|
| Brown Bear | 115 | 1,500 | 118,000 |
| Brown Bear | 225 | 3,000 | 181,000 |
| Cobey | 225 | 1,000–2,000 | 135,000–185,000 |
| Resource Recovery Systems | 300 | 2,000 | 104,000 |
| Resource Recovery Systems | 440 | 3,000 | 170,000 |
| Scarab | 234 | 2,000 | 104,000 |
| Scarab | 360 | 3,000 | 174,000 |
| Scat | 107 | 3,000 | 176,000 |

**Table 7.4 Partial List and Cost of Windrow Turning Equipment (PTO Driven)**

| Manufacturer | Power (HP) | Capacity (TPH) | Approximate Cost(U.S. $1991) |
|---|---|---|---|
| Centaur Walker | 90 | 800 | 7,400 |
| Scat | 65 | 2,000 | 55,000 |
| Wildcat | 70 | 1,000 | 46,500 |

the rain. In the absence of shelters, the possibility of the windrows taking in an excessive amount of moisture would be especially strong.

*Economics.* The economics of the "turned" type of composting varies according to the circumstances, size, and degree of mechanization of the process. As such, it has a wide variation and does not lend itself to ready generalization. Certainly the more complex operations are more expensive than those involving modest volumes of wastes that can easily be processed manually. On the other hand, windrow composting, whether it be static or turned, certainly is a much less expensive undertaking than is mechanized (enclosed) composting.

In a large-scale, turned windrow operation, probably the major expenditure for the actual composting (exclusive of size reduction and sorting) is for an automatic turner. The costs of representative machines were given in the preceding section. No attempt is made to estimate costs for erecting a shelter, because they vary extremely from one locale to another. Suffice it to state that the shelter need not be elaborate, although if the operation is to be conducted in the midst of a heavily populated area, provision should be made for deodorizing air vented from the shelter. It is difficult to arrive at firm costs on the basis of the exceedingly few reliable ones reported in the literature in the U.S. Those for manures and sewage sludge range from $30 to $60 per ton (1991).

*Limitations.* The major limitation of turned windrow systems probably is health oriented. Of course, the limitations are only applicable to operations that involve the processing of human excrement or residues from animals that harbor disease organisms pathogenic to man (zoonoses). The limitation stems from two features of turned windrows: (1) temperatures lethal to pathogens do not prevail throughout a windrow; in fact, toward the outer layers of a windrow, the temperatures may approach optimum levels for pathogens; and (2) the turning procedure may be instrumental in bringing about a recontamination of sterilized material by nonsterile material in the outer layers of the windrow in which bactericidal temperatures did not develop. (A temperature profile of a well-managed windrow is shown in Figure 7.3.) However, repeated turning eventually reduces the pathogen populations to less than infective concentrations. This latter condition is reached by the time the material is ready for final processing and use.

Turning improperly or insufficiently soon leads to a limitation in the form of generation of objectionable odors. Even with a suitable regimen, some odors are certain to be generated. However, the latter situation is a characteristic of any system that involves the handling and processing of wastes, whether it be static, turned windrow, or mechanized composting. The period in which odor generation is of nuisance proportions is during the preparatory and active stages of composting; hence, preventionary measures need be taken only during that time. The mere attainment of high temperatures does not betoken the absence of objectionable odors.

A lesser rapidity and resultant greater space requirement often have been alleged against the turned windrow approach as contrasted to the "high-speed" composting claimed for mechanical composting. The fallacy of this allegation was discussed earlier and is only reiterated at this point for the sake of emphasis. While on the subject of rapidity of composting, it should be emphasized that rapid composting becomes a virtue only when high-priced land area and costly reactor usage are involved. If no reactor is involved and land area is not critical, rapidity loses its advantage. Moreover, under those conditions, the intensity and frequency of turning can be lessened. The reason is that very little odor emanates from an intact pile of composting material. It is mainly during the turning process that foul odors, if present, are released from the pile. It should be emphasized, however, that this relaxation in terms of frequency and thoroughness is safe only when no human habitations are nearby, i.e., 500 or more feet away.

### *Mechanical Systems*

At first glance, the variety of types of reactors on the market seems to be very large. (In this section the term "reactor" is applied to the unit or complex of units in which the "active" stage of composting takes place.) The fact is that the breadth of the variety is more apparent than real, in as much as individually each type represents only a more or less slight departure from a few general characteristics. With few, if any, exceptions, they are designed to provide optimum aeration, and all do provide for the addition of moisture when needed. All employ one or more of the following: (1) forced aeration, (2) stirring, and (3) tumbling. Forced aeration usually is employed in systems in which the material is distributed as a layer in a trough. An exception is the Fairfield reactor. In this reactor, air is forced through the blades of augers. In some reactors, stirring is accomplished by distributing the material on a series of floors and rotating one or more plows through the material. In reactors in which aeration is through tumbling, the tumbling may be accomplished by dropping the material from one level to a lower level (from belt to belt or floor to floor), or by introducing it into a revolving drum, the interior of which is equipped with horizontally oriented vanes (e.g., Dano reactor).

A few randomly selected examples of systems on the market at the time of this writing are given in the paragraphs which follow. A more complete cataloguing and description can be found in Reference 19.

The Dano composter is not only typical of drum-type tumbling systems, it also is perhaps the most widely known of the various mechanical composting systems. The principal element in the Dano system is a long, almost horizontal drum, nine or more feet in diameter, that is rotated at about 2 RPM. The retention period in the drum is from 1 to 3 days. Material discharged after such brief detention periods is far from completely stabilized. Hence, specifications call for windrowing the material in which it is allowed to mature over a period of 1 month or longer. A recent application of the Dano drum in the U.S. incorporated manual segregation of inorganic matter before introduction of the waste into the reactor. In addition, aeration, curing, and upgrading stages were included.

A typical Dano installation is diagrammed in Figure 7.16. In 1991, the cost for a plant capable of processing 180,000 tons MSW per year was about $30,000,000. The cost of the plant includes manual separation and forced aeration.

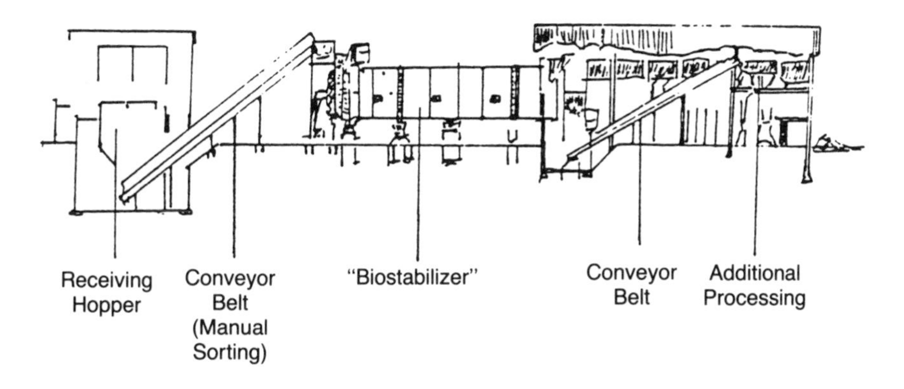


**Figure 7.16.** Typical Dano installation.

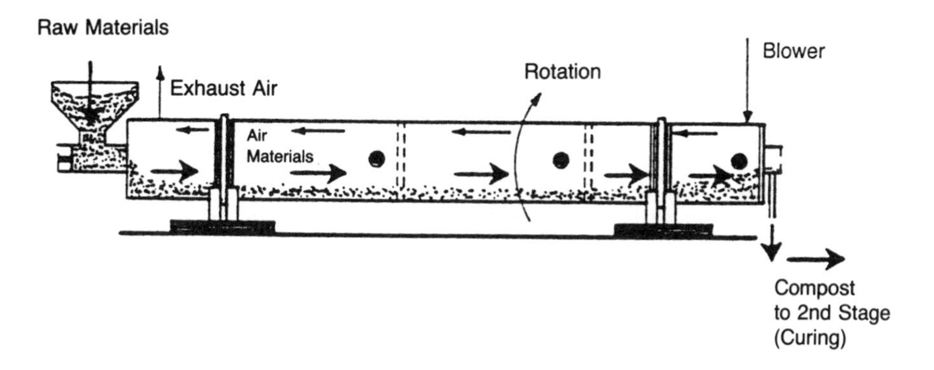


**Figure 7.17.** Rotating drum composter (Eweson).

Other versions of the drum type of reactor are the reactors used in the Ruthner system and the Eweson systems. A schematic diagram of the Eweson system is presented in Figure 7.17.

An example of a system that involved tumbling the material from one level to another is one originally known as the Naturizer System. Later, it was known as the Real Earth system. The system centered around the use of a plug-flow reactor housed in a vertical building equipped with horizontal, continuous conveyors on which the ground refuse was supposed to compost. The conveyors were about 9 ft wide and 150 ft long. The material was stacked to a depth of about 6 ft on the belts. At the end of a 24-hr period, the material on one belt was dropped or transferred to a lower belt. A system of fans provided the aeration. Usually after 2 days of exposure on the belts, the material was ground a second time and then reintroduced into the digester. In original versions, the detention time was set at 5 to 6 days. Later, the entrepreneurs spoke in terms of a 1- to 2-day detention period. Whether the period was 1 day or 6 days, the material was not sufficiently stabilized and was poor in quality. Consequently, a 5- to

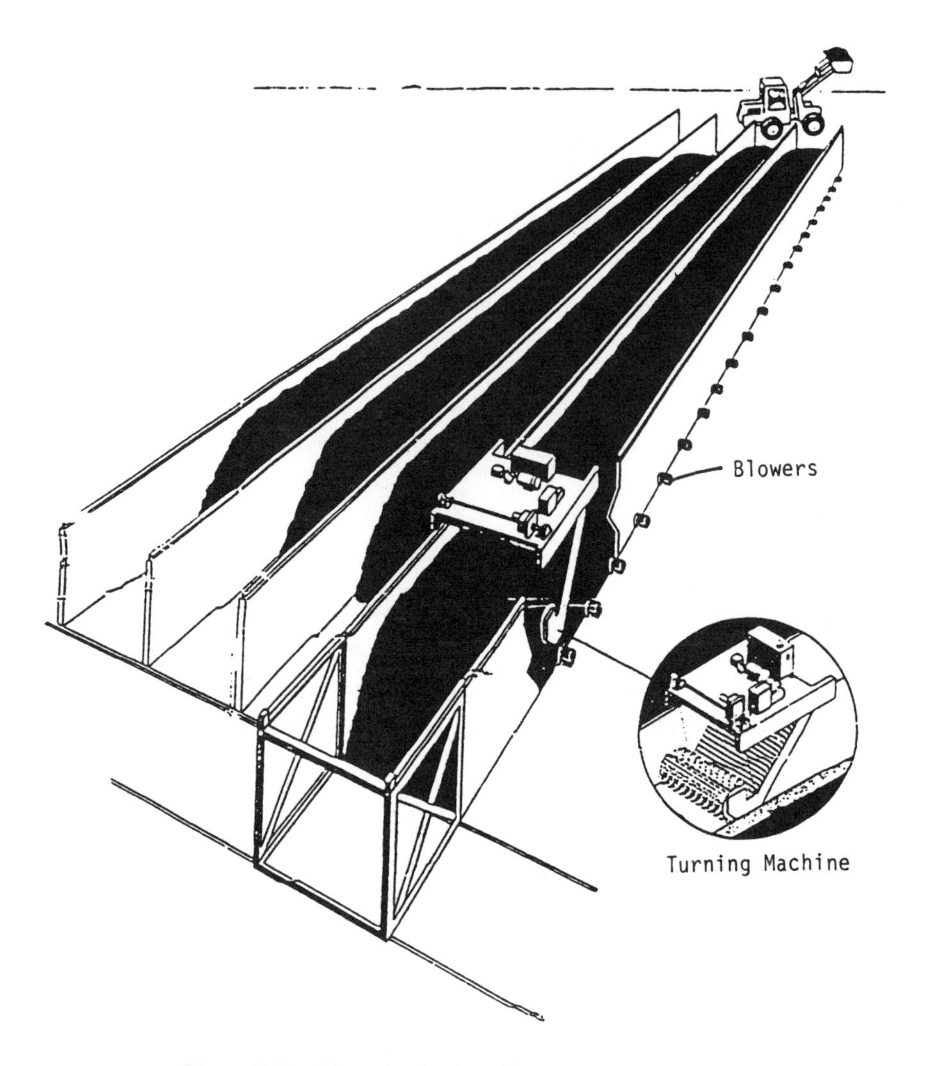


**Figure 7.18.** Example of agitated bed composting system.

6-week maturation period in windrows was added to the detention period in the reactor.

A system that combined forced aeration with tumbling was the Metro-Waste system. Tumbling was accomplished by means of a traveling endless belt. The forced aeration was accomplished by forcing air through the perforated floor of the trough or tank and into the composting mass placed on the floor. The reactor is diagrammed in Figure 7.18. The "agiloader" in the figure is the traveling endless belt. In operation, ground refuse was dumped into the troughs and air was forced through the composting mass on the order of one time each day. The detention period called for in promotional brochures was 6 days. Thereafter, the

**Figure 7.19.** Original Metro-Waste System.

material was windrowed and allowed to mature over a 1- to 2-month period. A photograph of the original system in Houston is presented in Figure 7.19.

In the authors' opinion, the forced aeration feature was an unnecessary adjunct in the Metro system, and aeration would have been better served by increasing the frequency at which the agiloader was passed through the material. Discarding the forced aeration feature would have made the Metro-Waste system a compromise between windrow and mechanized (confined) systems. As such, it would not have had the complexity of a mechanized systems and yet would have been more confined than "open" or windrow systems. Unlike many compromises that seem to lose the better features of the systems from which they were derived, those derived from Metro-Waste would very likely have several of the advantages of both the windrow and the mechanical systems. A system similar to the Metro-Waste system is presently being marketed by International Processing Systems (IPS) and by OTVD.

The Fairfield system exemplifies one that combines forced aeration and stirring. The Fairfield reactor (digester) consists of a cylindrical tank equipped with a set of screws ("augers" or "drills") supported by a bridge attached to a central pivoting structure. The reactor is diagrammed in Figure 7.20. The bridge with its collection of augers is slowly rotated. The augers are turned as the arm rotates. The augers are hollow and perforated at their edges. Air is discharged from the perforations and into the composting material. Detention time varies. If it is less than 2 or 3 weeks, the material must be windrowed in order to attain stability.

Another version of a forced air-stirring combination is that applied in the BAV (or BIAV) reactor, sometimes known as the silo system. The BAV reactor is diagrammed in Figure 7.21. It is a squat silo equipped with a screw especially designed to remove the "finished" product from the bottom of the reactor. Waste to be composted is fed through the top of the reactor and "finished"

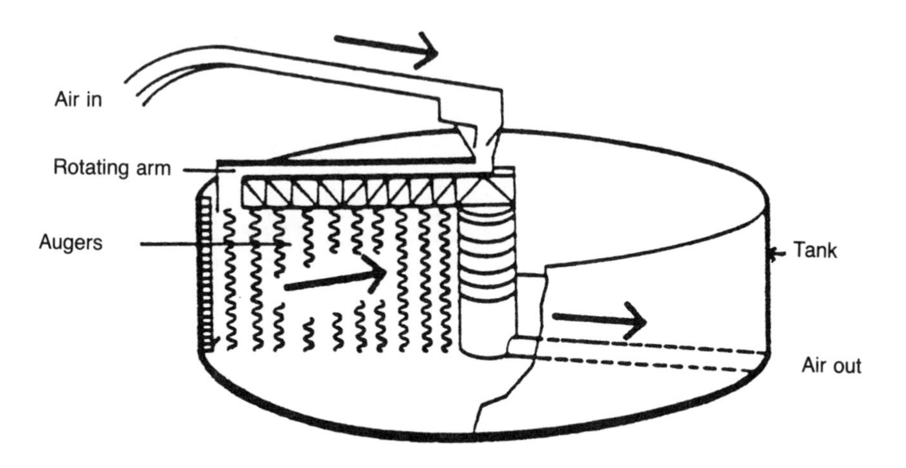


**Figure 7.20.** Diagram of the Fairfield reactor.

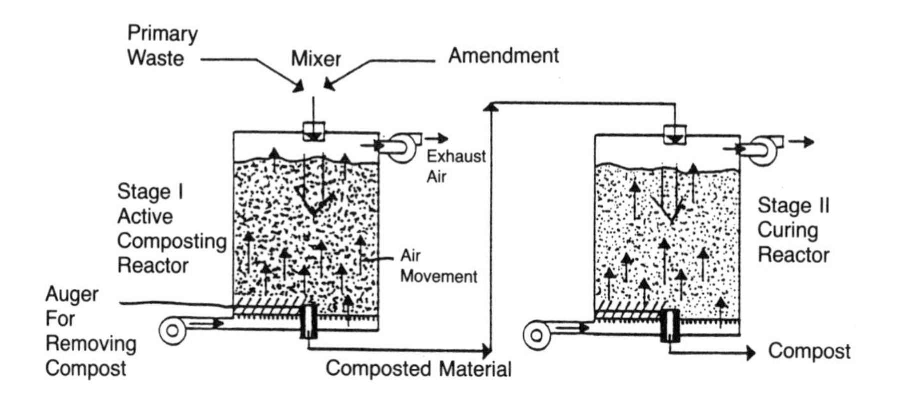


**Figure 7.21.** Silo composting system.

material is withdrawn from its bottom. Air is fed throughout the reactor. The bottom screw is helical in form and serves as an auger to move material to the center. As the screw rotates around the axis, the screw shaft moves radially about the center to bring in composting material from all around the circle. A problem frequently encountered in the operation of the unit is the tendency of the material "to bridge" over the screw. Detention times are on the order to 14 days. Thus far, most of the experience with the unit has been in the composting of sewage sludge.

*Economic Limitations.* The economics of some mechanized systems are more unfavorable than those of windrow systems. In the early 1970s, capital costs for compost plants in the U.S. were on the order of $15,000 to $20,000/ton of daily capacity. The operational costs were about $10 to $15/ton processed. Present costs (1991) range from about $15,000 to about $50,000/ton of daily capacity. Upon investigating the costs and the effort involved with a particular mechanized system, it should be borne in mind that a common failing in some of the promotional literature is the tendency to hold down apparent cost through the devices of underdesigning the equipment needed and underestimating operational requirements. Other factors to consider in making a comparative evaluation of a mechanized system were discussed in the section ''Technology — Principles.''

### Postprocessing

Postprocessing involves the various steps taken to refine the finished compost and meet regulatory and/or market requirements. Postprocessing may include one or more of the following unit processes: size reduction, screening, air classification, and destoning. These processes have been described in previous sections. In order to achieve adequate separation, the moisture content of the compost should be at or below 30%.

## IMPACTS OF THE OPERATION UPON THE ENVIRONMENT, PUBLIC HEALTH, AND INDUSTRIAL HEALTH

Principal potential negative impacts of a compost operation on the environment would be a lowering of the quality of water and air resources and the compromising of the public health and well-being by attracting and breeding vectors and rodents. It should be emphasized that these impacts are potential impacts and that they would become actual only when (1) an inadequate technology is used; (2) a normally adequate management is improperly applied; or (3) preventive or corrective measures are not taken. Moreover, environmental and public health hazards that might originate in a compost facility probably would be much less than those due to other forms of resource recovery from waste.

The impact on the land resource associated with the use and application of the compost product is discussed in Chapter 6.

### Water and Air Resources

#### *Water Resource*

Quality of the water resource can be adversely affected only through contamination with runoff from the compost operation or with leachate from raw, composting, or composted refuse.

Leachate is formed only when the moisture content of the material is excessively high (>60 to 65%). Aside from maintaining the moisture content of the material at or below permissible levels that approach optimum. Uncontrolled additions of moisture (e.g., rain, snow) can be minimized by sheltering the operation from the elements. As a precautionary measure, provision should be made to keep leachate from reaching ground and/or surface waters by conducting all phases of the operation on a suitably contoured, paved surface.

The paved surface should be equipped to collect all leachate for treatment or discharge into an on-site leachate treatment facility (lagoon) or into a public sewer. Runoff can be avoided by selecting a site where it would not be likely to occur. If this is not possible, runoff can be prevented from entering the operation site by excavating ditches to divert the runoff around the site. Runoff from the site can be intercepted and channeled to a treatment facility (e.g., conventional stabilization lagoon). Some techniques used to direct runoff from a composting site are presented in Figure 7.22. As shown in the figure, it is important that leachate does not reach a body of water, because leachate from raw waste is similar to that of raw sewage sludge in terms of pollutant concentration. Although the compost process sharply reduces the pollutant concentration, leachate from a properly matured compost would still reduce the quality of ground and surface waters.

### Air Resource

Biological and nonbiological agents from a compost operation most likely would enter the environment by way of dust particles and aerosols generated during the various stages of the operation and subsequently discharged into the air. Some of the microbes transported in this manner could be a hazard to the health of a susceptible individual who perchance might ingest the dust particle or aerosol. Aerosols are likely vehicles for a wide variety of microorganisms. These microorganisms may occur as single entities, as clumps of organisms, or as adhering to dust particles. Two types of infections may be acquired from such contaminants, namely, those that are limited to the respiratory tract and those that may affect another part of the body. Both are taken in by way of the respiratory tract.

The existence of a hazard from the spores of *Aspergillus fumigatus* is yet to be demonstrated. The infectivity of the spores is low. Consequently, any danger posed by it would be of significance only to an unusually susceptible individual. Nevertheless, prudence indicates that an open-air compost plant should not be sited in close proximity to human habitations.

Dust suppression at all stages of a compost operation can be accomplished through the use of conventional dust control measures. Some of the measures include the use of mist sprayers in the working area and the installation of air collection systems and particulate control devices such as cyclones and fabric filters.

*Odors.* Although the emission of objectionable odors lowers the quality of the air resource in terms of human well-being, it does not become a health hazard until the odors become particularly foul and intense.

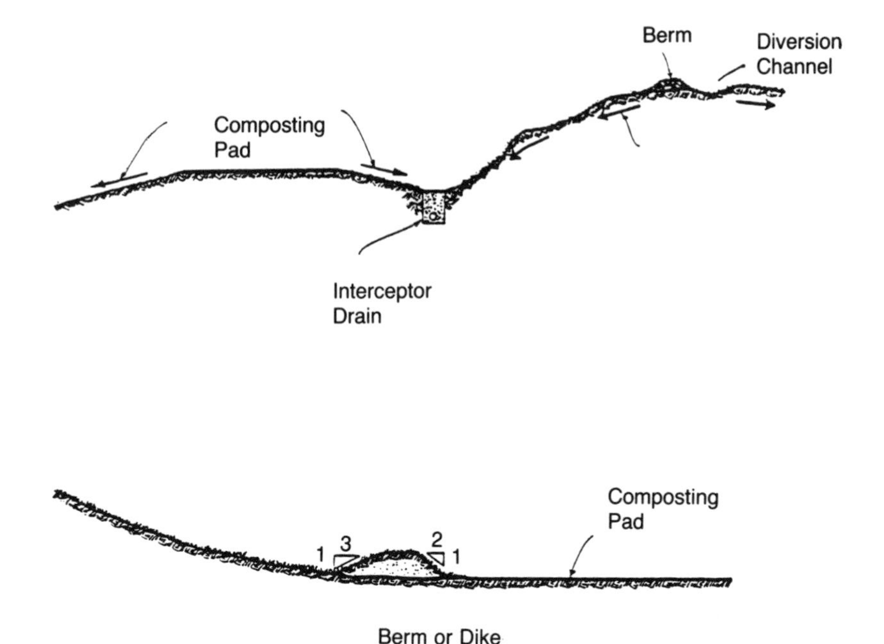


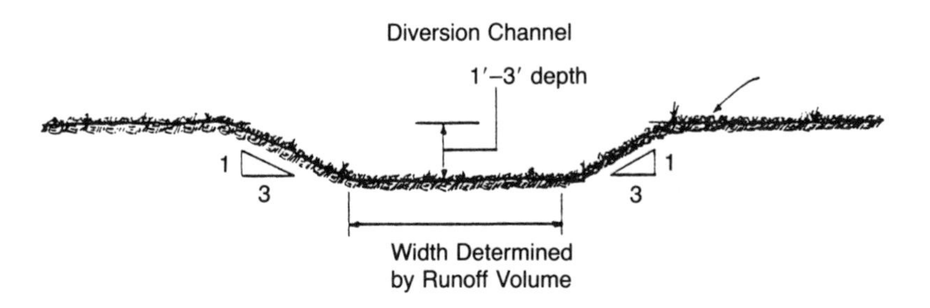


**Figure 7.22.** Run off diversion techniques (typical specifications).

Some odors are inescapable, e.g., those from raw wastes. Odor control during pre-processing can be accomplished by enclosing the entire operation in a building, conditioning the feed, and treating exhaust gases through absorption, adsorption, or oxidation methods. Foul odors are generated during the compost stages, principally through improper management of the composting process, e.g., failure to maintain adequate aerobic conditions. The use of an in-vessel system does not *ipso facto* guarantee the attainment of an odor-free operation.

In the absence of proper management, all composting materials not as yet fully matured can become sources of foul odors.

Conventional techniques are available for treating foul odors. Control and containment are effective approaches to preventing the odors from becoming problems, or of contending with those that escape prevention. Means of control involves trapping the odors through ventilation or containment of the compost process. Exhaust air can be treated by passing it through a chemical scrubbing system or by way of biofiltration. Difficulties with biofiltration are its large filter areas and relatively sophisticated management requirements.

### Vectors

With respect to composting refuse, food wastes, sewage sludge, or manures, fly and rodent attraction is almost inevitable. The main reasons for this are (1) the nature of the wastes and (2) the long time interval between reception of the raw material and the stages of the compost process in which conditions become lethal to flies and intolerable to rodents. Most likely, flies and rodents would not constitute a serious problems in a yard waste compost operation. However, food wastes or sewage sludge cocomposted with the yard wastes could serve as strong attractants for flies and rodents. Fly and rodent problems can be avoided to a considerable extent by enclosing the entire facility. In the absence of such containment, they can be minimized through the institution of certain measures. For example, an important mitigating measure would be careful "housekeeping" throughout all stages of the operation. Storage of raw waste should be as brief as possible. Preprocessing, particularly size reduction, accomplishes a substantial destruction of fly eggs and fly larvae and lowers the utility of the refuse as feedstuff for rodents. Migration of fly larvae from the site can be prevented by composting on a paved surface.

## INDUSTRIAL HEALTH AND SAFETY

Chances of incurring an injury through an accident are greatest during the preprocessing stage, because in this stage the exposure of workers to machinery (and in some cases raw wastes) is the most extensive. Standard measure for minimizing such hazards are well developed.

The greatest hazard to the health of the workers comes from the dusts suspended in the air in a compost plant. Hazards associated with dust are greatest in the preprocessing phase and become much less in the subsequent stages of the compost process. An additional biological burden generated by the dust particles and aerosols is the fibrous fraction of dust. The fraction may have a health significance. Health problems associated with dust can be substantially controlled through the use of face masks and protective clothing and the installation of adequate sanitation facilities.

Manual segregation of recyclable and nonprocessible items on the tipping floor is unsafe and should not be practiced without the utmost precautionary measures.

In general, industrial health and safety has been, thus far, ignored in most composting facilities.

The highest levels of noise that can occur in a preprocessing plant would be from about 95 dBA (slow response). These levels can be generated by a shredder or a front-end loader. With the present stage of the technology, some type of ear protection is needed for exposures longer than about 2 hr.

### Classification and Standards

Federal standards aimed at controlling the production and use of compost made from sewage sludge currently are available. Currently, there are not any federal regulations for the production and marketing of MSW compost. Some states have developed some standards. A listing of some of the standards is presented in Table 7.5. The data in the table show that some of the standards were adapted from sewage sludge and may not be applicable to other materials.

The authors participated in the development of a framework to institute standards for compost materials in the State of Washington.[20] The framework is based on the establishment of two sets of standards: (1) dealing with marketability and (2) addressing public health issues. The proposed standards are presented in Tables 7.6 and 7.7.

## HOME COMPOSTING

### Introduction

Because of the growing concern about resource conservation, shortages of suitable landfill sites, and diversion of wastes from the landfill, this section is devoted to composting by the householder at his or her residence. Home composting or back yard composting, if properly carried out, could divert an average of 20 to 30% of the residential waste stream from the landfill. In addition to conserving landfill space, many individuals also recognize the need to return to the soil the plant nutrients contained in MSW. This attitude may be due to an overall interest in the environment or in the protection of the ecology or it may be due to the individual's desire for an inexpensive and reliable source of compost for the garden.

The first major issue that faces the potential composter is the finding of reliable and comprehensible information on the subject. Many of the existing "recipes," booklets, and directions are, for the most part, the products of hearsay or folklore. The result is that, depending upon the particular document the individual may have read, the inexperienced composter is left either confused, discouraged, or overly optimistic. It is also possible that the individual may be persuaded to make the undertaking substantially more costly and complex than necessary. A second issue often met by the novice is the complaint of neighbors (particularly in high-density residential areas) that the compost pile stinks. The potential outcome of the complaints about foul odors is a decree from the local public health authorities either banning or restricting any further compost activities.

**Table 7.5 Compost Specifications for Various States in the U.S. (Parts per Million)**

| State:<br>Feedstock:[a] | FL<br>YD/MAN | FL<br>SL/MSW | FL<br>SL/MSW | FL<br>SL/MSW | MA<br>SL/MSW<br>YD | MA<br>SL/MSW<br>YD | MN | NY<br>SL/MSW | PA<br>SL/MSW |
|---|---|---|---|---|---|---|---|---|---|
| Usage:[b]1/ | U 2/ | U | L2 | L4 | U | L1 | U | U | L3 |
| Criteria | | | | | | | | | |
| Mercury | — | — | — | — | 10 | 10 | 5 | 10 | 10 |
| Cadmium | 15 | 15 | 100 | >100 | 2 | 25 | 10 | 10 | 25 |
| Molybdenum | — | — | — | — | 10 | 10 | — | — | — |
| Nickel | 50 | 50 | 500 | >500 | 200 | 200 | 100 | 200 | 200 |
| Lead | 500 | 500 | 1,500 | >1,500 | 300 | 1,000 | 500 | 250 | 1,000 |
| Chromium | — | — | — | — | 1,000 | 1,000 | 1,000 | 1,000 | 1,000 |
| Copper | 450 | 450 | 3,000 | >3,000 | 1,000 | 1,000 | 500 | 1,000 | 1,000 |
| Boron | — | — | — | — | 300 | 300 | — | — | — |
| Zinc | 900 | 900 | 10,000 | >10,000 | 2,500 | 2,500 | 1,000 | 2,500 | 2,500 |
| PCB | — | — | — | — | 2 | 10 | 1 | 1 | 3 |
| Particle size (mm) | <25 | <10 | <25 | <25 | — | — | — | <10 | — |
| Foreign material | <10% | <10% | <30% | <40% | — | — | — | — | — |
| Maturity | — | Mature | Mature or Semi | Can be fresh | Stable | Stable | Mature | — | — |

[a] YD = yard debris; MAN = manure; SL = sludge; MSW = Municipal solid waste.

[b] 1/ usage: U = unrestricted distribution; L = limited distribution: (1) non-food chain crops; (2) commercial, agricultural, institutional, or governmental agencies; (3) public distribution; (4) landfill or land reclamation uses. 2/ not subjected to testing; compost is assumed to meet limits for heavy metals.

**Table 7.6 Proposed Marketability Related Standards**

| | Unit | Grade A | Grade B |
|---|---|---|---|
| Bulk density | lb/cu yd | 600–800 | 400–1000 |
| CEC | meq/100 g | >100 | >100 |
| Foreign matter | Maximum % | 2 | 5 |
| Moisture content | % | 40–60 | 30–70 |
| Odor | | Earthy | Minimal |
| Organic matter | Minimum % | 50 | 40 |
| pH | | 5.5–6.5 | 5–8 |
| Size distribution | Nominal-in. | <1/2 | <7/8 |
| Water holding capacity | Minimum % | 150 | 100 |
| C:N | Maximum | 15 | 20 |
| Nitrogen | Minimum % | 1 | 0.5 |
| Conductivity (soluble salts) | mmhos/cm | <2 | <3 |
| Seed germination | Minimum % | 95 | 90 |
| Viable weed seeds | | None | None |

*Source:* State of Washington Department of Ecology.[20]

**Table 7.7 Proposed Public Health Related Standards**

| | Unit | Class 1 | Class 2 | Class 3 |
|---|---|---|---|---|
| Arsenic | ppm | 10 | 15 | 20 |
| Cadmium | ppm | 5 | 10 | 50 |
| Chromium | ppm | 100 | 200 | 1000 |
| Copper | ppm | 300 | 500 | 1000 |
| Lead | ppm | 250 | 500 | 1000 |
| Mercury | ppm | 3 | 5 | 10 |
| Nickel | ppm | 50 | 100 | 200 |
| Zinc | ppm | 500 | 1000 | 2000 |
| Organochlorine pesticides | ppm | ND | a | a |
| Organophosphorous pesticides | ppm | ND | a | a |
| PCBs | ppm | ND | 1 | 5 |
| PCPs | ppm | ND | 1 | 5 |
| Fecal streptococci[a] | MPN/100 ml | | | |
| Fecal coliform[a] | MPN/100 ml | | | |

*Source:* State of Washington Department of Ecology.[20]

[a] Limits to be determined.

This section of our book is written with the hope of providing the required information to avert the pitfalls mentioned in the preceding paragraph. It will also give the reader suggestions or clues as to the conduct of individual experimentation to find the method best suited to a particular situation.

## General Requirements

Certain general requirements must be satisfied in order to conduct a successful composting operation at home. First of all, the material to be composted (known as raw material, feedstock, or substrate) must have the proper composition and must be in a suitable physical condition. For an efficient operation, a minimum volume of raw material must be processed. In the method to be presented in this section, a bin or some type of container is needed in order to keep the composting material within a confined space. In addition, a set of steps or procedures must be followed in setting up and conducting the compost operation.

The remainder of this section deals with descriptions and discussions of these major requirements.

### *Feedstock*

As previously indicated, composting is a biological process. Consequently, the process is limited by the constraints that are characteristic of any biological activity. One of the basic requirements in any biological activity is the availability of essential nutrients. As pointed out in the preceding sections, the essential nutrients include carbon, nitrogen, phosphorous, and trace elements. These nutrients not only must be present, but they also must be present in a form available to the microogranisms involved in composting. In other words, they must be present as compounds that the organism can use.

Specifically, appropriate materials for composting are garden trimmings, manures, garbage, vegetable trimmings, paper and cardboard, and various absorbent materials. In other words, any decomposable organic material is suitable. Exceptions for reasons of public health are human feces, diseased animals, plant debris heavily dosed with pesticides, and toxic material in general.

### *C:N*

A nutritionally related requirement generally unknown to the nonmicrobiologist is that carbon and nitrogen must be present in certain proportions. The limiting ratio is usually the maximum one — in this case from 25:1 to 30:1; i.e., 25 to 30 parts of carbon to 1 of nitrogen. Too high a C:N slows the process; whereas too low a C:N leads to a nitrogen loss in the form of ammonia. The reasons were given in an earlier section.

For home composting, the C:N requirement can be met by adjusting the proportion of "green" debris or of garbage to "dry" garden debris. Examples of green garden debris are lawn clippings, green leaves, green plant stems, roots, flowers, etc. Dry debris refers to dried (no longer green) grass (hay), matured flower stalks, branches (excluding leaves), straw, etc. The green material is rich in nitrogen — and hence increasing the ratio of "greens" to "drys" lowers the use of manures. Poultry manure is the most effective because of its high nitrogen content. Pet feces would also be good sources of nitrogen.

One may wonder how the C:N of the raw material can be determined. Unfortunately, determining the C:N involves the conduct of some exacting and relatively expensive tests. Consequently, some trial and error coupled with good judgment are necessary. A useful way to assure an adequate ratio is to follow the old Indore method in setting up the pile, i.e., layering. Dry layers are alternated with green or moist layers. Each layer is from 2 to 4 in. thick.

The dry material is used for absorbing the excess moisture, as well as to impart structural strength to the pile. It keeps the porosity and prevents compaction. Observation tells us that sawdust, straw, and dried leaves are ideal materials for the purpose. Paper is not useful as an absorbent because it becomes soggy and compacts.

Other materials that are useful or even required, are phosphorous, and an array of trace elements. Generally these substances are present in sufficient quantity in plant debris and manures.

Some publications on home composting call for the addition of lime to keep the pile from becoming too acid in its reaction. In scientific terminology, this means that the lime is added to raise the pH of the pile — i.e., makes it turn from acid to alkaline. The reasoning is that most microorganisms cannot thrive under acid conditions or may even be killed. Actually, the ideal pH level for most microorganisms is 7.0, a level that corresponds to neutrality. At pH 7.0, a material is neither acid nor alkaline. These authors have found that while most piles become somewhat acid during the onset of decomposition, this condition was not detrimental nor did it last for a very long period of time. The trouble with adding lime is that it promotes the loss of nitrogen. The reasons are straightforward. The lime brings about a rise in pH to alkaline levels. At alkaline levels, the ammonium radical leaves its ionized state and is volatilized — it becomes a gas. Combining this volatilization with the high temperatures characteristic of an actively composting mass leads to an extensive loss of nitrogen in the form of ammonia. It must be admitted, however, that the addition of lime is followed by an improvement in the physical appearance and an ease in handling of the composting material.

### *Minimum Volume*

Under the conditions prevailing in the San Francisco Bay Area (California), the minimum recommended volume is 1 cu yd. There is nothing of magic about the dimension of 1 cu yd. It simply is the minimum volume at which a pile becomes sufficiently self-insulating to retain its heat. Undoubtedly, in the Dakotas in midwinter, a 3 cu yd pile may prove to be too small. The cubic yard volume is a convenient one for home composting because the amount is neither too unwieldy nor overwhelming, nor so tiny as to seem ''piddling'' to the gardener.

The home composter need not wait until he or she has accumulated a cubic yard of wastes from his own grounds. He can borrow debris from his neighbors, obtain vegetable trimmings from the local supermarket, or purchase a couple of sacks of steer manure. The manure not only provides a source of nitrogen, it also serves as an excellent absorbent.

If the volume is built up through accumulation, then care should be taken to turn the material occasionally in the manner to be described later. Otherwise, odor and fly problems may be encountered.

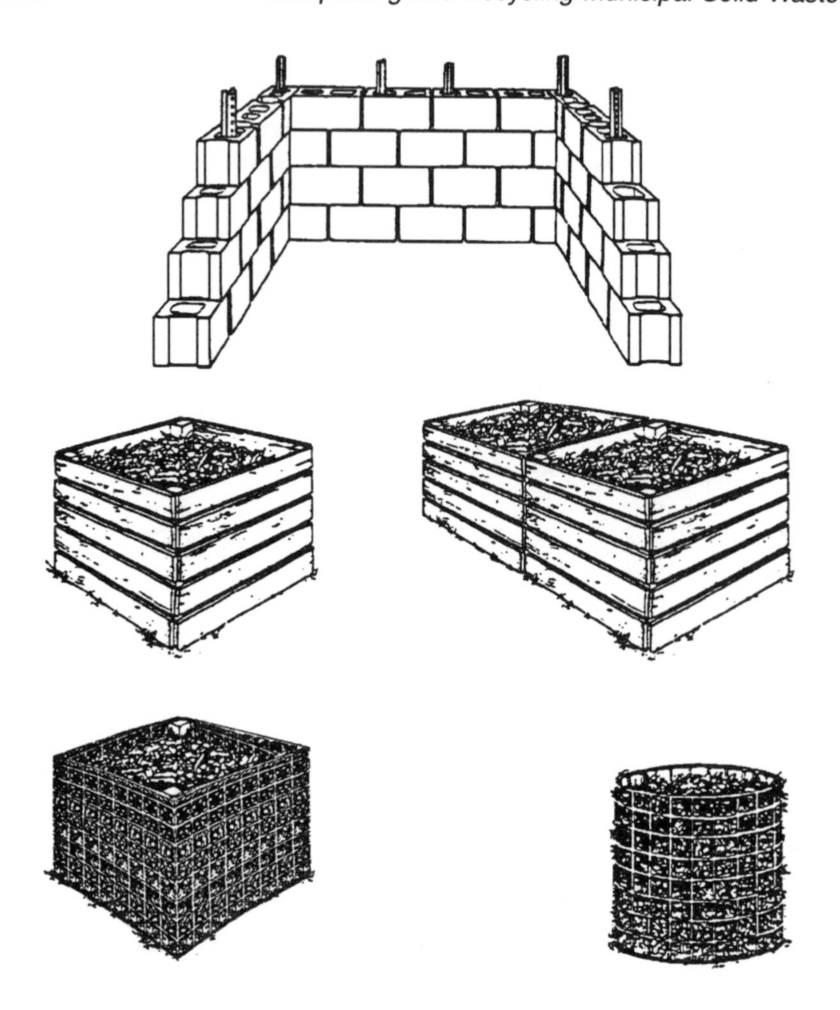

**Figure 7.23.** Examples of bins for home composting.

### *The Bin*

The bin may be constructed of concrete, wood, or even of hardware cloth, as shown in Figure 7.23. The minimum floor dimensions should be 3 x 3 ft. The height can be from 4 to 6 ft. It should be constructed such that one side can be removed to provide easy access to the composting material. If hardware cloth is used, the mesh size should be 1/4 in. and the wire should be of heavy gauge. Obviously, the cloth should be attached to a sturdy frame. An advantage in the use of hardware cloth is that all surfaces of the composting material are exposed to air. Disadvantages are (1) the same exposure becomes a severe detriment if a fly problem should arise; (2) the life span of such a bin will not be as long as that of a concrete or wooden bin; and (3) in a cold climate or in a windy area, little protection is had against heat loss.

A workplace equal to two to three times that of the floor space of the bin should be provided in front of the bin. This space is needed for turning the material. Another consideration is that of constructing a double bin. The twin bins can have a common wall, i.e., in effect a double-sized bin divided into two by a center partition. Material could be accumulated in one bin, while material in the second bin is composting.

To allay the qualms of local health authorities, the top surface (the exposed surface) should be covered with a fly-proof screen.

## Procedure

### *Preparation of Raw Material*

If manure, or sawdust, or leaves, or dry grass is used as the absorbent, and vegetable trimmings, garden debris, lawn trimmings, and kitchen garbage as the nitrogen source, very little preparation of raw material is needed. For ease of handling, it would be advisable to chop up long flower stalks (e.g., dahlias, marigolds, etc.) and vegetable stalks (corn, kale, pea vines, etc.) into pieces about 6 or 8 in. long. The chopping can be done with a sharp spade. Straw should be chopped into 3- or 4-in. pieces. While shredding the material with a power shredder would be ideal, it is not absolutely necessary. In studies conducted by one of the authors on ''backyard'' composting, excellent results were obtained without recourse to shredding.

As stated earlier, when starting a compost run, the material should be layered in the bin such that absorbent layers alternate with ''wet'' layers. The height of the pile should not exceed 6 ft. If the pile is too high, the material tends to compact and thereby exclude the necessary air from between the particles.

### *Moisture Content*

Generally, if vegetable trimmings, lawn clippings, or kitchen garbage are used, the initial moisture content of the material will be adequate for composting. A good rule of thumb to follow is that the material is sufficiently moist if the surface of the particles glistens. An insufficiency of moisture will be manifested by the failure or delay of the material to ''heat-up.'' An excess of moisture is soon revealed by the development of a foul odor and a drop in heating. The moisture content can be raised by sprinkling with tap water. A condition of excessive moisture can be remedied by adding more absorbent material or by increasing the frequency of turning to once a day.

### *Turning (Aeration)*

In as much as the reasons for turning the material were given in an earlier section, only the technique is given at this time.

To begin the turning procedure, the front of the bin is removed. Then the contents of the bin are taken out, beginning with the top. Care should be taken to keep track of the position of the material in the bin so that when it is returned, that material which constituted the outer layers of the mass end up in the interior of the reconstituted pile. In short, the turning should be such that every particle in the original pile is at some time in the interior of the pile. When transferring the material to and from the pile, it should be "fluffed" to insure maximum porosity in the pile. During the course of the first turning and thereafter, materials in the absorbent and wet layers should be mixed. In other words, the layering of wet and dry materials is discontinued with the first turning.

The authors have found a five-tined pitchfork to be an excellent tool for turning.

Frequency of turning is a function of the composter's ambition, moisture content of the pile, and the urgency with which the finished compost is desired. The more frequent the turning (up to once a day), the faster the material composts. One of the authors was able to "compost" a mixture of dry leaves (sycamore), vegetable trimmings, and garden debris (total volume — 1 cu yd) in 12 days under the following schedule: (1) first turning on the third day after starting the pile; (2) second turning on the third day after the first turning (i.e., skip a day); and (3) a third and final turning on the ninth day after setting up the pile. Composting was complete by the 12th day.

### *Monitoring the Process*

The best way to monitor the process is by noting the course of the temperature. This can be done by purchasing a hotbed thermometer and using it to check the temperature inside the pile. The temperature inside the pile (about 12 to 15 in. from the surface) should rise to 110 to 120°F within 24 to 48 hr after starting the process. It should reach 130°F and higher within 3 or 4 days after the start. Thereafter the temperature will remain at 130°F or higher until all of the readily decomposable material is stabilized. Then, the temperature will drop. When it drops to around 110°F, the material is ready for use.

### *Trouble-Shooting*

If the temperature does not rise, or if it drops suddenly, the pile may be too wet, too dry, or the C:N is too high. If the moisture content is too high, as stated earlier, the material will stink. If it is too low, the material will have a dry appearance. If the C:N is too high no odor will be noticeable, and the material will glisten if the moisture content is satisfactory. The remedies for each of these problems were given in the preceding paragraphs.

### *Indore Method*

For the less ambitious, the Indore method of composting is the suitable one for home composting. The Indore method derives its name from the region of India in which it was developed. Composting according to this method involves a minimum of effort. In this procedure, layers of straw or dry, coarse material are alternated with manure or green organic matter. The mixture is placed either in open piles or in pits. The composting mass may be turned or left undisturbed. Turning the mass hastens decomposition. In either case the material should be covered with a 2-in. layer of thoroughly compacted soil to prevent fly production and reduce odors. If the mass is turned, the first turning should be 8 to 10 days after placing the material. The second should occur 30 to 40 days thereafter. The compost should be ready for use about a month later. If not turned, composting usually requires a year. Most farm and garden journal methods are modifications of the Indore process. The process is almost entirely anaerobic and therefore does not generate sufficient heat to kill undesirable organisms and seeds that may be present.

Composting garden and kitchen waste according to the Indore method is best done in pits or bins. If space permits, a pit some 5 x 3 x 2 ft deep may be dug in the garden area. The pit is then filled with layers of lawn clippings, weeds, and garden refuse. This may be done gradually, taking care to cover each fresh addition with a layer of soil. Once the pit is filled, it should remain undisturbed until the following year at which time the humus may be removed and used. If a tightly constructed bin is used, the same procedure holds with the exception that it should be approximately 5 ft high. In addition to covering with soil, it is advisable to keep the bin tightly screened on top to prevent access of flies to the interior of the bin.

Treatment of farm manure and farm waste products according to the Indore method is similar to that given for garden refuse, but using a larger pit or bin. Open piles or windrows may be used when large volumes are involved, but they should be immediately covered with a 2-in. tightly compacted layer of soil to prevent fly production.

## REFERENCES

1. "Second Interim Report of the Interdepartmental Committee on Utilization of Organic Wastes," *N. Z. Eng.*, 61–12: November–December (1951).
2. "Composting Fruit and Vegetable Refuse: Part II," in *Investigations of Composting as a Means for Disposal of Fruit Waste Solids*, Progress Report, National Canners Assoc. Research Foundation, Washington, DC. August (1964).
3. Schulze, K.F. "Rate of Oxygen Consumption and Respiratory Quotients during the Aerobic Composting of Synthetic Garbage," *Compost Sci.*, 1:36, Spring (1960).
4. Schulze, K.F. "Relationship Between Moisture Content and Activity of Finished Compost," *Compost Sci.*, 2:32, Summer (1964).
5. Chrometzka, P. "Determination of the Oxygen Requirements of Maturing Composts," *International Research Group on Refuse Disposal, Information Bulletin 33,* August (1968).
6 Lossin, R.D. "Compost Studies: Part III. Measurement of the Chemical Oxygen Demand of Compost," *Compost Sci.*, 12:12–31 March-April (1971).

7. Regan, R.W. and J.S. Jeris. "A Review of the Decomposition of Cellulose and Refuse," *Compost Sci.*, 11:17, January-February (1970).
8. Senn, C.L. "Role of Composting in Waste Utilization," *Compost Sci.*, 15(4):24–28, September-October (1974).
9. Golueke, C.G. *Composting* (Emmaus, PA: Rodale Press, 1972).
10. Niese, G. "Experiments to Determine the Degree of Decomposition of Refuse by Its Self-Heating Capability," *International Research Group on Refuse Disposal, Information Bulletin 17*, May (1963).
11. Rolle, G. and E. Orsanic. "A New Method of Determining Decomposable and Resistant Organic Matter in Refuse and Refuse Compost," *International Research Group on Refuse Disposal, Information Bulletin 21*, August (1964).
12. Möller, F. "Oxidation-Reduction Potential and Hygienic State of Compost from Urban Refuse," *International Research Group on Refuse Disposal, Information Bulletin 32*, August (1968).
13. Obrist, W. "Enzymatic Activity and Degradation of Matter in Refuse, Digestion: Suggested New Method for Microbiological Determination of the Degree of Digestion," *International Research Group on Refuse Disposal, Bulletin 24*, September (1965).
14. Lossin, R.D. "Compost Studies," *Compost Sci.*, 11:16, November-December (1970).
15. Jourdan, B. "Determination of the Degree of Decompositions in Waste and Waste/Sludge and Compost," *Stuttgart Berichte Zur. Abfallurtschaft*," Vol. 30, 1988.
16. Wylie, J.S. "Progress Report on High-Rate Composting Studies," *Engineering Bulletin, Proc. of the 12th Industrial Waste Conf.*, Series No. 94, May (1957).
17. Epstein, E., G.B. Willson, W.D. Burge, D.C. Mullen, and N.K. Enkiri, "A Forced Aeration System for Composting Wastewater Sludge," *J. Water Pollut. Cont. Fed.*, 48(4):688, April (1976).
18. Golueke, C.G. and P.H. McGauhey. *Reclamation of Municipal Refuse by Composting*, Tech. Bulletin No. 9, Sanitary Engineering Research Laboratory, University of California, Berkeley, June (1953).
19. Dean, R.B. "European Manufacturers Display Systems at Kompost '77," *Compost Sci.*, 19(2):18–22 March-April (1978).
20. *Compost Classification/Quality Standards for the State of Washington*, prepared by Cal Recovery Systems, Inc., in association with Wilsey & Ham Pacific, Inc., Dr. Charles L. Henry, and Thomas/Wright, Inc., for the State of Washington Department of Ecology, September (1990).

# CHAPTER 8

# Products and Markets

Marketing plays a critical role in determining the feasibility of a program designed to recover resources from urban wastes. Securing markets for recycled materials accomplishes two important objectives — it provides an end use for the finished product and it provides a source of revenue which could partially offset the cost of processing and contribute to the financial viability of an operation.

Because of the differences in the types of markets for compost and for recyclables (e.g., paper, plastic, ferrous), as well as the strategies and issues involved, the two categories of recycled materials are discussed separately in this chapter.

## COMPOST

Compost has the potential to be used as a soil amendment for many different applications. When incorporated into the soil, compost increases the organic content of the soil and can improve its texture, nutrient content, and water retention and aeration capacities. Compost can also be used as a mulch, as a top dressing, or as landfill cover. The various uses for compost and its benefits to the soil are discussed in Chapters 6 and 7.

Due to the beneficial characteristics of compost, the product has several potential applications and can be used by a variety of market segments. These include:

- Landscaping
- Land reclamation (landfills, quarries, etc.)
- Erosion control
- Top dressing (e.g., for golf courses, park land)
- Agriculture
- Residential gardening
- Nurseries

Each potential application is limited by the qualities of the product and the constraints applicable to its use.

## Potential Markets

The potential size of the compost market largely depends upon the quality of the compost product and the types of uses for the material. Composts from different types of wastes (e.g., mixed MSW, yard waste, food waste, source-separated MSW) have different characteristics and therefore have different potential markets.

### *MSW Compost*

The agriculture industry is the largest potential market for compost, particularly MSW compost. The reason is twofold: (1) the nature and quality of MSW compost; and (2) the immense quantities of product that would be available if composting were widely adopted as a means of treating mixed MSW. Land reclamation and improvement rank next in potential size of market.

Studies have shown that the sustained application of compost has beneficial effects that include favorable soil pH, higher crop yields, increased organic matter, increased cation exchange capacity, enhanced supply of plant nutrients, and increased water retention.[1]

Certain recent developments in agriculture have broadened the recognition of the usefulness of compost. One of the developments is an increasing understanding of the serious consequences of topsoil erosion, particularly the reduction in soil fertility.[2] Topsoil erosion is due to a large extent to the dwindling concentration of organic matter in soil. The realization that the lost organic matter can be replaced by an organic soil amendment such as compost is stimulating the market for compost and other organic soil amendments.[3]

Another beneficial effect that can aid in developing the market for soil amendments in the agriculture market is the reduction of solubility of inorganic fertilizer elements (NPK) brought about by the action of organic matter. Conversion of highly soluble inorganic fertilizer elements to slowly soluble organic forms has the net effect of increasing efficiency of nutrient use by crop plants.

In recent years, the competition between compost and chemical fertilizers has begun to diminish, and the two products are beginning to be viewed as being complementary rather than competitive. This shift in attitude is due to the growing recognition of the importance of the following four facts:[2,3] (1) the presence of organic matter in soil increases the efficiency of chemical fertilizer utilization in crop production; (2) compost can serve as a supplementary rather than competing source of NPK; (3) compost enriches the organic matter content of soil and thereby enhances the water retention capacity of the soil; and (4) the cost of chemical fertilizers is increasing.

Nevertheless, agriculture remains a difficult market to penetrate. Some of the problems that need to be overcome to develop the market are availability of compost at the appropriate time, consistency in composition and nutrient content, ensuring low levels of toxic substances, difficulties of bulk application, and acceptance by farmers.

A classification of agriculture that is useful for evaluating the market potential of the compost product is a breakdown into field crop, row crop, orchard, and ornamental categories.

*Field Crops.* Because of the large amount of land area involved, it would be expected that field crop agriculture would provide the major market for MSW compost. However, due to difficulties in application in comparison with competing products, the actual market size is relatively small. MSW compost competes poorly with chemical fertilizers because its bulk density and NPK content are much lower than those of chemical fertilizers. Bulk densities of MSW compost range from 700 to 1000 lb/cu yd and, consequently, large quantities of the product must be handled. Thus, to apply a given amount of NPK to a field crop, the amount of chemical fertilizer is in terms of pounds per acre, whereas with compost it would be in tons per acre. Many farmers regard the large amounts involved as a distinct disadvantage because of the associated inflation of costs of transport and application.

The reluctance of field crop agriculture to use compost has been aggravated by the unavailability of a product having a consistent quality. Unavailability is a negative factor because a market cannot be developed without a reliable and sufficiently large supply of a product with consistent specifications.

*Row Crops, Orchards, Ornamentals.* The market for compost in row crop and orchard agriculture and ornamentals has not reached its full potential, largely due to a limited supply of good quality compost. This unfortunate situation inevitably will change with the closing of landfills, the implementation of nonburn alternatives, and the diversion of the various organic residues (e.g., sewage sludge, leaves, yard debris) from the landfill to composting facilities.

*Land Reclamation.* The land reclamation market for compost differs from the agriculture market in that land reclamation usually is a responsibility of the government, whereas agriculture usually is privately controlled. Hence, purchasing compost and other soil amendments for land reclamation is done by the government. Regardless of whether the government produces the compost or purchases it from a private producer, the potential demand always will be large. The economic aspects of the market depend on the willingness of the government to pay for the compost required.

### *Yard Waste Compost*

Results of a limited number of marketing surveys for composted yard waste may be found in the open literature. Market evaluations conducted by the authors show that the organizations or individuals who typically express the most interest in using yard waste compost are

- Landscape contractors and suppliers
- Sod and sodding services

- Retailers of soil conditioners
- Nurseries
- Public agencies
- Home gardeners

Generally, landscapers, sod and sodding services, and retailers of soil amendments buy compost or other soil amendments in bulk. Most businesses express preferences for certain specifications for the compost product. The need for consistency in the quality of products is frequently mentioned. Other specifications of interest are nutrient content; possible contamination with pathogens, heavy metals, and toxic compounds; appearance of the compost; and purchasing convenience.

Public agencies have the potential to use both high- and low-quality composts. High-quality compost can be used in areas where humans and/or animals may come in contact with the materials. A lower quality compost would be suitable for land reclamation and as landfill cover. Examples of potential uses by public agencies include parks and redevelopment, weed abatement on public lands, land upgrade, roadway maintenance, and median strip landscaping.

The residential segment also represents a substantial market for soil amendments. Results of marketing surveys conducted by the authors and their colleagues show that many individuals express a preference to use compost from yard waste rather than from MSW or sewage sludge. This preference largely stems from concerns related to disease transmission, contamination, and safety. In order to successfully market compost to the residential sector, the public needs to be assured of product safety and informed on the uses of compost.

## Product Quality

Marketing studies conducted throughout the U.S. have identified quality and consistency of the product as key elements in the utility and marketability of the products. Assuming a properly conducted composting operation, compost quality is ultimately determined by (1) the composition of the feedstock; (2) type and thoroughness of the separation process; and (3) adequacy of the composting process.

There is an ongoing trend towards improving compost quality through the control of feedstocks and the process. Compost that is produced from source separated organic materials (e.g., yard waste) typically has a higher and more consistent quality than that produced from mixed MSW, due to a more homogeneous feedstock and a smaller concentration of contaminants.

### *Compost Quality Parameters*

Compost quality is a function of its physical, chemical, organic, and biological characteristics. A list of parameters that may be used to define the quality of compost is presented in Table 8.1. As shown in the table, the various parameters define different aspects of quality, e.g., aesthetics, handling, and health.

**TABLE 8.1 Parameters That May Be Used to Define the Quality of Compost**

| Category | Parameter | Purpose |
|---|---|---|
| Physical | Bulk density | Transportation, handling (e.g., application), storage |
| | Color | Aesthetics |
| | Moisture content | Handling |
| | Odor | Health, environment, aesthetics, marketability |
| | Organic matter content | Soil quality |
| | pH | Soil quality |
| | Size distribution | Handling, aesthetics, soil quality |
| | Water holding capacity | Soil quality, water conservation |
| | Contaminants (inert matter) | Public and animal health, soil quality, environment, aesthetics |
| | Maturity | Soil quality, crop production, stability |
| Chemical | Nutrients (macro and micro) | Soil quality, crop production |
| | Heavy metals | Soil quality, health, environment |
| | Soluble salts | Soil quality, crop production, environment |
| Organic | Toxic compounds | Health, environment |
| Biological | Pathogens | Health, environment |
| | Seed germination | Soil quality |
| | Weed seeds | Soil quality, crop production |

*Source:* Washington Department of Ecology.[5]

Among the physical characteristics usually desired of a product are dark color, uniform particle size, earthy odor, absence of contaminants, moisture content, nutrient content, and organic matter.

*Color.* The composting process generally results in a darkening in color as the process advances. For example, the organic fraction of MSW generally changes from a grayish green color to black as it undergoes the composting process. A deep, dark material is usually associated with stability, maturity, and a high concentration of organic matter.

*Particle Size.* The size distribution of the compost defines the utility of the material. A uniform size distribution is necessary, for example, for use by container nurseries, whereas a less uniform size distribution may be acceptable for field crop agriculture or erosion control. A relatively small particle size also enhances eye appeal and promotes ease of application. The size distribution of the individual particles defines the texture of the product and therefore affects the productivity of the soil. Texture determines porosity, permeability, and other parameters that are important for plant production.

*Odor.* Odor is an imprecise but effective means of monitoring a composting process. The presence of foul odors is usually an indication of anaerobic conditions. On the other hand, the presence of a strong earthy odor is a good indication of a mature compost.

*Contaminants.* The acceptable level of contamination in a compost product is set by the intended use for the material. The compost should not contain identifiable contaminants such as glass shards and pieces of metal and plastics. Neither should it contain harmful concentrations of toxic compounds, pathogens, or weed seeds.

*Moisture Content.* Moisture content is an important parameter with regard to ease of handling. To facilitate transport and application, the moisture content of the product should be lower than 50%.

*Nutrient Content.* Although the nutrient content of MSW compost typically is low and is not a true indicator of the utility and value of the product, the nutrients in compost have the advantage of being in a form that can be used by plants. Major plant nutrients are nitrogen, phosphorous, and potassium. Minor nutrients include copper, manganese, iron, and boron.

*Organic Matter.* The real utility of compost is its ability to increase the humus content of soil. Humus enhances the friability and water holding capacity of soil.

### *Compost Characteristics*

As stated earlier, the characteristics of a given compost product are dependent upon the composition of the feedstock and the steps used to process the material. The approximate nitrogen content and carbon to nitrogen ratios of a broad list of compostable feedstocks are presented in Table 8.2. The data demonstrate that the C:N ratio of these materials ranges from as low as 0.8 for urine to 511 for raw sawdust. Consequently, feedstocks to a compost operation should be combined in such a way as to ensure the production of a finished compost product with an acceptable C:N ratio.

Data are provided in Table 8.3 on the physical, chemical, organic, and biological characteristics of composts produced from yard waste, wastepaper, food waste, and mixed MSW. As shown in the table, there is considerable variation both among the types of composts and within each type of compost.

### *Compost Classification Standards*

The establishment of a formal set of standards and specifications for monitoring the quality of compost is gaining momentum in the U.S. Standards were first introduced in the late 1970s due to a growing interest in composting sewage sludge and were intended primarily to protect public health and safety. Consequently, the parameters regulated most often were heavy metals and pathogens.

It was later recognized that the standards needed to be extended to include other municipal wastes. The states of Minnesota and New York introduced solid waste compost quality standards in 1988. Since that time, a number of other

**TABLE 8.2 Approximate Nitrogen Content and C:N Ratios of Some Compostable Materials (Dry Basis)**

| Material | Nitrogen Content (%) | Carbon-to-Nitrogen Ratio |
|---|---|---|
| Urine | 15–18 | 0.8 |
| Blood | 10–14 | 3 |
| Fish scrap | 6.5–10 | — |
| Poultry manure | 6.3 | — |
| Mixed slaughterhouse waste | 7–10 | 2 |
| Night soil | 5.5–6.5 | 6–10 |
| Activated sludge | 5.0–6.0 | 6 |
| Meat scraps | 5.1 | — |
| Purslane | 4.5 | 8 |
| Young grass clippings | 4.0 | 12 |
| Sheep manure | 3.75 | — |
| Pig manure | 3.75 | — |
| Amaranthus | 3.6 | 11 |
| Lettuce | 3.7 | — |
| Cabbage | 3.6 | 12 |
| Tomato | 3.3 | 12 |
| Tobacco | 3.0 | 13 |
| Onion | 2.65 | 15 |
| Pepper | 2.6 | 15 |
| Cocksfoot | 2.55 | 19 |
| Lucerne | 2.4–3.0 | 16–20 |
| Kentucky blue grass | 2.4 | 19 |
| Grass clippings (average mixed) | 2.4 | 19 |
| Horse manure | 2.3 | — |
| Turnip tops | 2.3 | 19 |
| Buttercup | 2.2 | 23 |
| Raw garbage | 2.15 | 25 |
| Ragwort | 2.15 | 21 |
| Farmyard manure (average) | 2.15 | 14 |
| Bread | 2.10 | — |
| Seaweed | 1.9 | 19 |
| Red clover | 1.8 | 27 |
| Cow manure | 1.7 | — |
| Wheat flour | 1.7 | — |
| Whole carrot | 1.6 | 27 |
| Mustard | 1.5 | 26 |
| Potato tops | 1.5 | 25 |
| Fern | 1.15 | 43 |
| Combined refuse, Berkeley, CA (average) | 1.05 | 34 |
| Oat straw | 1.05 | 48 |
| Whole turnip | 1.0 | 44 |
| Flax waste (phormium) | 0.95 | 58 |
| Timothy | 0.85 | 58 |
| Brown top | 0.85 | 55 |
| Wheat straw | 0.3 | 128 |
| Rotten sawdust | 0.25 | 208 |
| Raw sawdust | 0.11 | 511 |
| Bread wrapper | nil | — |
| Newspapaer | nil | — |
| Kraft paper | nil | — |

*Source:* Gotaas.[4]

**TABLE 8.3 Characteristics of Different Types of Compost**

| | Unit[a] | Yard Waste | Wastepaper[b] | Food Waste[c] | Mixed MSW |
|---|---|---|---|---|---|
| Physical | | | | | |
| Bulk density | lb/yd$^3$ | 515–680 | 550[a] | 600[a] | 600–800[a] |
| CEC | mmhos | 0.8–4.07 | | | |
| Moisture content | % | 34–61 | ~40 | ~40 | ~45 |
| pH | | 5.8–7.2 | | | |
| Size distribution | Nominal in. | 1.1–4.0[d] | 1–4[e] | <3/4[e] | 1–4[e] |
| Water holding capacity | % | 110–300 | | | |
| Chemical | | | | | |
| Aluminum | ppm | 600 | | | |
| Arsenic | ppm | 5.0 | 54.6 | 1.8 | 1.1–9 |
| Boron | ppm | 0.5–81 | 24.4 | 3.8 | |
| Cadmium | ppm | 0.8 | 1.2 | 0–0.8 | 1.4–6.0 |
| Calcium | ppm | 875 | 2,800 | 2,511–38,400 | |
| Carbon (total) | % | 32.5 | 17.8 | 7.3 | |
| Chlorine | ppm | 192 | | | |
| Chromium | ppm | 23 | 27.1 | 6.6–23.8 | 16.2–220 |
| Cobalt | ppm | | 7.0 | 1.8–17.2 | |
| Copper | ppm | 2–6 | 61.6 | 8.7–24.8 | 46.5–630 |
| Cyanide | ppm | | | | 0.49 |
| Iron | ppm | 144–412 | 2,945 | 922 | |
| Kjeldahl nitrogen | % | 0.7 | 0.53 | 0.95 | |
| Lead | ppm | 72 | 82.4 | 7.5–23.8 | 82.4–913 |
| Magnesium | ppm | 11–920 | 3,100 | 1,050–9,000 | |
| Manganese | ppm | 36 | 64.2 | 21–181.6 | |
| Mercury | ppm | 0.06 | 0.02 | 0.01 | 1.2–5.0 |
| Molybdenum | ppm | | 16.3 | 9.2 | |
| Nickel | ppm | 22 | 18.4 | 2.1–2.9 | 8.3–110 |
| Nitrates | ppm | 2–8 | | | |
| Phosphorous | ppm | 4–280 | 640 | 1,300–4,600 | |
| Potassium | ppm | 184–3,600 | 3,100 | 5,800–8,900 | |
| Sodium | ppm | 39–753 | | | |
| Soluble Salts | mmhos | 0.7–1.9 | | | |
| Zinc | ppm | 19–160 | 142 | 24.7–110.2 | 266–1,650 |
| Organic | | | | | |
| 2,4,5-T | ppm | <0.5 | | | |
| Aldrin | ppm | 0.007 | | | ND–0.07 |
| Casoron | ppm | Present | | | ND |
| Chlordane | ppm | 0.15–0.32 | | | ND |

| | | | |
|---|---|---|---|
| Dalapon | ppm | <0.5 | |
| Diazinon | ppm | ND | ND |
| Dicamba | ppm | 0.5–12.9 | |
| Dichloroprop | ppm | <0.5–1.2 | |
| Dieldrin | ppm | 0.019 | 0.04–0.08 |
| Dinoseb | ppm | <0.5–1.0 | |
| Dursban | ppm | 0.039 | ND |
| Endrin | ppm | ND | ND |
| Lindane | ppm | ND | 0.08 |
| Malathion | ppm | ND | ND |
| MCPA | ppm | <0.5–1.2 | |
| MCPD | ppm | <0.5–7.1 | |
| o p DDT | ppm | 0.004 | |
| p p DDE | ppm | 0.005–0.014 | ND |
| p p DDT | ppm | 0.008–0.019 | ND |
| Parathion | ppm | ND | ND |
| PCBs | ppm | ND | 0.32–2.53 |
| PCPs | ppm | 0.12–0.21 | 0.016 |
| Silvex | ppm | <0.5 | |
| Trifluralin | ppm | Present | ND |
| Biological | | | |
| *A. fumigatus* | | Negative | Negative |
| *Ascaris lumbr.* | | Negative | Negative |
| Coliform (total) | MPN/100 ml | $1.4 \times 10^3$–$300 \times 10^3$ | $4.0 \times 10^8$ |
| Dog parasitic ova | | Negative | Negative |
| *E. coli* | MPN/100 ml | $<1.0 \times 10^4$ | $>1.1 \times 10^8$ |
| Fecal coliform | MPN/100 ml | $2.3 \times 10^3$–$93 \times 10^3$ | $>1.1 \times 10^8$ |
| Human parasitic ova | | Negative | Negative |
| *Pseudomonas* spp. | | Positive | Positive |
| *Salmonella* sp. | | Negative | Negative |
| *Taenia* sp. | | Negative | Negative |
| Trichuris trichuria (hookworm) | | Negative | Negative |

[a] Dry weight basis.
[b] Calculated.
[c] Limited available data supplemented by calculated values.
[d] Visual.
[e] Estimated.

*Source:* Washington Department of Ecology.[5]

**TABLE 8.4 Examples of State Regulations for Compost**

| | New York | Minnesota | Florida | Maine |
|---|---|---|---|---|
| **Classification** | **Class I** | **Class I** | **Code 1** | **Class A** |
| **Feedstocks** | **Yard Debris MSW/Sludge** | **MSW** | **Yard Debris MSW/Sludge** | **MSW Sludge** |
| Heavy metals (ppm) | | | | |
| Cadmium | 10 | 10 | <15 | 10 |
| Lead | 250 | 500 | <500 | 700 |
| Mercury | 10 | 5 | | 10 |
| Zinc | 2500 | 1000 | <900 | 2000 |
| Chromium | 1000 | 1000 | | 1000 |
| Nickel | 200 | 100 | <50 | 200 |
| Copper | 1000 | 500 | <450 | 1000 |
| Organic compounds | | | | |
| PCBs (ppm) | <1 | <1 | | <10 |
| Dioxin (ppt) | | | | |
| Food | | | | <27 |
| Nonfood | | | | <27–250 |
| Physical Properties | | | | |
| Moisture content (%) | | 50–60 | | |
| Odor | | Control | | |
| pH | | | 5.0–8.0 | |
| Particle size (mm) | <10 | <25 | <10 | |
| Biological Properties | | | | |
| Human pathogens | | Minimized | | |

*Source:* Eggerth and Diaz.[7]

states, including Florida, Maine, New Hampshire, North Carolina, and Washington, have either implemented standards or are in the process of doing so. The solid waste compost standards generally have been patterned after sludge standards and have regulated yard waste compost less stringently than MSW compost.

Examples of state regulations for compost are presented in Table 8.4. A few states have developed a classification hierarchy in which compost products meeting the most stringent requirements are allowed the widest distribution, i.e., nonrestricted use. Conversely, composts with higher levels of contaminants are restricted in use to non-food chain crops, to use as landfill cover, or to land reclamation projects.

At present, classification development is rapidly changing. Table 8.5 is included to indicate the current situation in Europe and to allow a comparison between the requirements in the U.S. and those in Western Europe.

Another change taking place in the U.S. is the interest in developing compost quality standards that are not necessarily related to public health and safety. This interest is prompted by the realization that the marketability of the product can be enhanced by ensuring a consistently high-quality product with known characteristics. These characteristics pertain to the utility rather than the safety of the product, such as moisture content, particle size, and carbon to nitrogen ratio. Table 8.6 presents a grading scheme being considered by the state of Washington.[5] The grading scheme would be voluntary. However, a facility would be allowed to classify its compost as a "Grade A" material only if the specifications were met. The producer would be required to provide the user with the specifications for the particular grade of compost being marketed.

**TABLE 8.5 Council of European Communities (CEC) Physical and Chemical Parameters for Compost**

| | Recommended | Mandatory |
|---|---|---|
| Heavy metals (ppm, dry wt) | | |
| Cadmium | 5 | 5 |
| Lead | 750 | 1000 |
| Mercury | 5 | 5 |
| Zinc | 1000 | 1500 |
| Chromium | 150 | 200 |
| Nickel | 50 | 100 |
| Copper | 300 | 500 |
| Organic matter (% dry wt) | | >30 |
| Particle size (mm) | | <24 |
| Minimum detention (days) | | Variable |
| Moisture content | | Variable |
| Conductivity (grams salt per liter) | | <2 |
| pH | | 5.5–8.0 |
| Maximum inerts (% dry wt) | | |
| Glass | | 4 |
| Plastics | | 1.6 |
| Minimum mineral content (% dry wt) | | |
| N | | 0.6 |
| $P_2O_5$ | | 0.5 |
| $K_2O$ | | 0.3 |
| CaO | | 2.0 |
| $CaCO_3$ | | 3.0 |
| MgO | | 0.3 |
| Carbon-to-Nitrogen Ratio | | <22 |

*Source:* Washington Department of Ecology.[5]

**TABLE 8.6 Example of a Voluntary Grading Scheme for Compost**

| | Unit | Grade A | Grade B |
|---|---|---|---|
| Bulk density | lb/cu yd | 600–800 | 400–1000 |
| CEC | meq/100 g | >100 | >100 |
| Foreign matter | Maximum % | 2 | 5 |
| Moisture content | % | 40–60 | 30–70 |
| Odor | | Earthy | Minimal |
| Organic matter | Minimum % | 50 | 40 |
| pH | | 5.5–6.5 | 5–8 |
| Size distribution | Nominal in. | <1/2 | <7/8 |
| Water holding capacity | Minimum % | 150 | 100 |
| C:N | Maximum | 15 | 20 |
| Nitrogen | Minimum % | 1 | 0.5 |
| Conductivity (soluble salts) | mmhos/cm | <2 | <3 |
| Seed germination | Minimum % | 95 | 90 |
| Viable weed seeds | | None | None |

*Source:* Washington Department of Ecology.[5]

## Marketing Issues

### *Purchase Considerations*

Compost products usually are sold in bulk and in bagged form. Preferences are heavily dependent upon the quantities used. Large-scale users (e.g., landscapers, nurseries, farmers, park districts) generally prefer to buy compost in bulk, whereas small-scale users (e.g., home gardeners) often prefer the bagged form. Sizes of bagged compost usually range from 1 to 2 cu ft. Bagged compost frequently is purchased through nurseries and garden supply stores.

Most MSW and yard waste compost is currently marketed in bulk form at no cost or a relatively low cost. The major portion of MSW compost currently is given away free; some of the compost is sold for $4.50 to $10/cu yd. High-quality yard waste compost sells, in bulk, for between $14 and $16/cu yd. In some instances, the yard waste compost is bagged and sold for the equivalent of $30/cu yd or more (1992 prices[7]).

### Competing Products

The size of the entire soil amendment market varies substantially from community to community. The type and quantity of each of the marketed amendments depend upon the availability of soil amendment product and the type of industry in each location.

A number of soil amendment materials are considered to compete with compost produced from mixed MSW and yard waste. However, as shown in Table 8.7, in many cases these products may actually be complementary.

Competing products include soils (e.g., fill dirt, topsoil), wood products (e.g., wood chips, bark, bark mulch, sawdust), and other organic soil amendments (e.g., peat, potting soil, mushroom compost, sewage sludge compost, animal manure, manure compost, straw). Many of the soil amendment products used and sold are blends of two or more materials.

*Soils.* The types and characteristics of soils sold nationwide depend upon the predominant soil type in the locale and the expected use of the soil. Soils are typically required by the construction industry and by landscapers to increase the elevation of an area, to minimize erosion, as a growth medium, or as fill material.

Based on a nationwide survey conducted in 1989,[8] the prices for fill dirt range from $2.50 to $11/cu yd; topsoil, from $4 to $17.50/cu yd; and screened topsoil, from $11.50 to $18/cu yd.

*Wood Products.* Wood chips, bark, bark mulch, and sawdust are used extensively by landscapers and homeowners to conserve moisture, control weeds, and for decorative purposes. Based upon a nationwide survey, the use of wood chips by office park developers and residents is becoming increasingly popular and is an important retail market in certain regions of the country.[8] The 1989 survey reported that the prices for bark mulch ranged from $12 to $25/cu yd and that wood chips sold for $12 to $30/cu yd.

*Other Organic Soil Amendments.* Other organic soil amendments that compete with MSW and yard waste composts include potting soil, peat, mushroom compost, sewage sludge compost, animal manure, manure compost, and straw.

**TABLE 8.7 Competing Products with Compost**

| Material | Degree of Competition[a] | | Basis of Competition | Use as Complement |
|---|---|---|---|---|
| | Compete | Complement | | |
| Top soil | X | XX | Organic matter<br>Porosity<br>Moisture retention | Mixed with top soil for specific applications |
| Fill dirt | | XX | — | Compost placed on top of fill dirt |
| Bark mulch and wood chips | X | | Moisture retention<br>Weed control<br>Erosion prevention | |
| Potting soils | X | XX | Moisture retention<br>Porosity<br>Organic matter | Mixed for special needs |
| Manures | XX | X | Porosity<br>Organic matter | May be mixed with manure |
| Peat | X | XX | Moisture retention<br>Porosity | Mixed with peat as special potting blend |

*Source:* U.S. EPA.[8]

[a] XX indicates one particular use.

A variety of potting soils are commercially available using different mixtures of soils, peat, clay, compost, perlite, sand, sawdust, vermiculite, and vermicompost. The price for potting soil depends upon the composition of the mixture, the grade of materials used, the manufacturer, and its form (e.g., bagged or bulk). A nationwide market analysis reported that prices for potting soils ranged from $2.59 to $3.99 for a 40-lb bag; and from $20 to $30/cu yd in bulk.[8]

One of the most important features of peat moss is its capacity to absorb and retain water while at the same time maintain adequate quantities of oxygen. In addition, it is free of weeds, diseases, and pests and is readily penetrated by plant roots.[9] Peat moss is used extensively by horticulturists, greenhouse operators, landscapers, and homeowners. It is sold in bales or bags for $2 to $5.50/cu ft.[8]

### *Constraints on Use of Compost*

A low-quality compost can have an adverse effect on human health and safety; animal health and safety; crop production; and the quality of the air, water, and land resources. Consequently, constraints on the use of the product may be necessary in order to avoid or minimize the adverse effect. The importance of the individual constraints depends upon whom or what is affected and the extent to which they are affected.

*Human Health and Safety.* Constraints on the use of the compost with respect to the health and safety of humans arise from harmful substances that may be within the product. Examples of such substances are pathogenic organisms, heavy metals, PCBs and other toxic organic compounds, and glass. The waste used as feedstock for the compost process is the source of harmful substances that may appear in the product.

The harmful effect on humans and animals may be exerted directly by eating food crops grown on soil that had been amended with compost, or indirectly through the consumption of meat and other products involving animals fed on such crops. The effects are due to the persistence of inorganic contaminants and the survival of certain pathogens through the food chain.

Cadmium can be used to demonstrate how a heavy metal passes through the food chain. A certain amount of the cadmium in soil is assimilated by plants growing on the soil. The amount assimilated by the plant depends upon several factors. Among the factors are the availability of the metal, plant species, and the particular part of the plant (e.g., root, leaf, fruit). Soil characteristics that affect availability of a particular metal are concentrations of the metal, pH, concentration of organic matter, ion exchange capacity, and several other characteristics. Regarding soil pH, availability usually decreases as the soil pH becomes less acidic and more alkaline. As a rule, leafy vegetables assimilate more than do cereal crops. In cereal crops, concentration is greater in the root and leafy portions than in the grain. If those plants are eaten by humans, a fraction of the cadmium or other metal in the plants is assimilated in the tissues of the persons who eat the plants. If the plants are consumed by animals, a

**TABLE 8.8 Concentration of Total Metals in Compost (Acid Digestion, in ppm)**

| Compound | MSW (a) | Sewage Sludge (b) | Sewage Sludge (a) | Yard Debris (a) | Yard Debris (a) |
|---|---|---|---|---|---|
| Cyanide | 0.49 | | | 0.15 | 0.08 |
| Mercury | 3.70 | | | 0.05 | 0.08 |
| Arsenic | 1.10 | | | 4.80 | 5.20 |
| Cadmium | 4.80 | 26.23 | 19.60 | 0.80 | 0.80 |
| Chromium | 56 | 288.47 | 543 | 24.20 | 21.60 |
| Nickel | 32.80 | 126.02 | 148 | 21 | 22.70 |
| Lead | 913 | 384.01 | 423 | 72.90 | 71.50 |
| Magnesium | 5,800 | | 2,300 | 2,500 | 2,600 |
| Calcium | 76,000 | | 9,900 | 10,500 | 10,300 |
| Sodium | 4,700 | | 300 | 200 | 200 |
| Iron | 13,200 | | 13,700 | 13,500 | 15,000 |
| Aluminum | 5,400 | | 11,300 | 7,800 | 7,000 |
| Manganese | 340 | | 338 | 396 | 3,390 |
| Copper | 190 | 485.12 | 560 | 25 | 42 |
| Zinc | 1,010 | 1,368.02 | 1,400 | 160 | 160 |

*Source:* Portland Metropolitan Service District.[14]

*Note:* (a) One sample; (b) average of 10 samples.

certain fraction of each metal is assimilated by the animals and remains in their meat and in products (e.g., eggs) produced by them. The size of the fraction depends upon the identity of the metal. Finally, individuals who consume the meat or products of the animals assimilate a fraction of the metals in the meat and products. The distribution of cadmium in the soil, plant, and animal as it passes through the food chain are described in detail in References 10 to 13. The incidence of other metals and chemicals in the food chain are summarized in Reference 14.

*Types of Constraints.* Because in the past the composting of wastes usually involved the use of sewage sludge as the feedstock, much of the data on the presence of harmful substances in compost pertain to sludge compost. Data from a study conducted in Portland, OR, on the characteristics of MSW, yard waste, and sewage sludge composts are presented in Tables 8.8 to 8.10.[14] As shown in the tables, concentrations of heavy metals and pathogens were reported to be higher in compost produced from sludge than in compost produced from the organic fraction of municipal solid wastes or from yard wastes. A summary of the Portland compost comparison is presented in Table 8.11.

When considering the data, it should be kept in mind that concentrations vary widely from operation to operation due to a variety of site-specific differences (e.g., residential MSW vs industrial MSW).

Toxic Substances. Available data on MSW compost indicate a wide range of values for concentrations of heavy metals. As shown in Table 8.3, the concentrations for chromium ranged from 16.2 to 220 ppm; copper, from 46.5 to 630 ppm; lead, from 82.4 to 913 ppm; and zinc, from 266 to 1650 ppm. In some instances, relatively high concentrations of PCBs are beginning to appear in MSW compost.[5,14] However, the reports are too fragmentary to indicate trends.

**TABLE 8.9 Concentration of Herbicides, Pesticides, PCBs and PCPs in Compost (in ppm)**

| | Municipal Solid Waste | | Sewage Sludge | Yard Debris | |
|---|---|---|---|---|---|
| Herbicide/pesticide | | | | | |
| Chlordane | ND[a] | ND | ND | 0.324 | 0.152 |
| p′p′DDE | ND | ND | ND | 0.014 | 0.005 |
| p′p′DDT | ND | ND | ND | 0.019 | 0.008 |
| o′p′DDT | ND | ND | ND | 0.004 | ND |
| Toxaphene | ND | ND | ND | 0.300 | 0.300 |
| Aldrin | ND | ND | ND | ND | 0.007 |
| Dieldrin | ND | ND | ND | ND | 0.019 |
| Dursban | ND | ND | ND | ND | 0.039 |
| Endrin | ND | ND | ND | ND | ND |
| Lindane | ND | ND | ND | ND | ND |
| Malathion | ND | ND | ND | ND | ND |
| Parathion | ND | ND | ND | ND | ND |
| Diazinon | ND | ND | ND | ND | ND |
| Trifluralin | ND | ND | ND | Present | Present |
| Casoron | ND | ND | ND | Present | Present |
| Alpha-BHC | ND | ND | | | |
| Beta-BHC | ND | ND | | | |
| Delta-BHC | ND | ND | | | |
| Gamma-BHC | ND | ND | | | |
| 4,4′-DDD | ND | ND | | | |

| | | | | | |
|---|---|---|---|---|---|
| 4,4′-DDE | ND | ND | | | |
| 4,4′-DDT | ND | ND | | | |
| Endosulfan I | ND | ND | | | |
| Endosulfan II | ND | ND | | | |
| Endosulfan sulfate | ND | ND | | | |
| Endrin aldehyde | ND | ND | | | |
| Heptachlor | ND | ND | | | |
| Heptachlor epoxide | ND | ND | | | |
| Dalapon | | | | <0.50 | <0.50 |
| Dicamba | | | | 0.5–12.9 | <0.50 |
| MCPD | | | | <0.5 | <0.50 |
| MCPA | | | | 0.5–7.1 | <0.5–2.4 |
| Dichloprop | | | | <0.5 | <0.5–1.2 |
| 2,4-D | | | | <0.5 | <0.5 |
| Silvex | | | | <0.5 | <0.5 |
| 2,4,5-T | | | | <0.5 | <0.5 |
| 2,4-DB | | | | <0.5 | <0.5 |
| Dinoseb | | | | <0.5–1.0 | <0.5–1.0 |
| Polychlorinated biphenyls (PCBs) | 2.53 | 2.34 | 0.52 | ND | ND |
| Pentachlorophenol (PCPs) | 0.016 | | 0.015 | 0.210 | 0.120 |

*Source:* Portland Metropolitan Service District.[14]

[a] ND = not detected/below detection limit. Blank spaces indicate tests were not conducted.

**TABLE 8.10 Concentration of Pathogens Found in the Various Types of Composts**

| Test | MSW | Sewage Sludge | Yard Debris | |
|---|---|---|---|---|
| Salmonella | Neg | Neg | Neg | Neg |
| *E. coli* | $>1.1 \times 10^8$ | $>1.1 \times 10^8$ | $>1.0 \times 10^3$ | $<1.0 \times 10^4$ |
| Fecal coliform | $>1.1 \times 10^8$ | $>1.1 \times 10^8$ | $2.3 \times 10^3$ | $9.3 \times 10^4$ |
| Total coliform | $4.0 \times 10^8$ | $4.3 \times 10^7$ | $1.4 \times 10^3$ | $3.0 \times 10^5$ |
| *Aspergillus fumigatus* | Neg[a] | Neg | Neg[b] | Neg[b] |
| Human parasitic ova | Neg | Neg | Neg | Neg |
| Dog parasitic ova | Neg | Neg | Neg | Neg |
| *Entamoeba coli* | Neg | Neg | Neg | Neg |
| *Entamoeba histolytica* | Neg | Neg | Neg | Neg |
| *Pseudomonas* spp. | Positive | Neg | Positive | Positive |
| *Ascaris lumbriocoides* (roundworm) | Neg | Neg | Neg | Neg |
| *Taenia* spp. (tapeworm) | Neg | Neg | Neg | Neg |
| *Trichuris trichuria* (hookworm) | Neg | Neg | Neg | Neg |

*Source:* Portland Metropolitan Service District.[14]

[a] *Rhizopus* species found
[b] *Rhizopus* and *Geotrichum* found.

TABLE 8.11 Comparison Between Three Types of Composts

| Characteristic | Yard Debris Compost | Sewage Sludge Compost | MSW Compost |
|---|---|---|---|
| Appearance | Satisfactory; granular shape; relatively uniform particle size | Good, fine granular quality (based on visual inspection) | Fair; some inert matter (based on visual inspection) |
| Heavy metals | Some data missing; most likely not a limitation on use of product or application rate | Relatively high concentrations; not permitted for food crop production, including animal feedstuff production | Higher concentrations than yard debris compost; not a restrictive factor |
| Salt concentration | Within maximum limit for most plants | Within maximum limit for most plants | Relatively high concentration |
| Weed/crop seeds | None | None | None |
| Germination | Favorable | Not favorable | Not favorable |
| Herbicides/pesticides | Most below detection limits | None detected | Most below detection limits |
| PCBs | None detected | Low levels detected | High concentration detected |
| Nutrient content | Low in nitrogen | Good | Fair |

*Source:* Eggerth et al.[6]

Results of analyses on yard waste compost (see Table 8.3) indicate that heavy metals concentrations are much lower than for MSW compost: 23 ppm of chromium, 2 to 6 ppm of copper, 72 ppm of lead, and 19 to 160 ppm of zinc. Concentrations of pesticides and PCBs are at or below limits of detection.[5,14]

Pathogens. Constraints due to the presence of pathogens in compost range from negligible to substantial, depending upon the waste composted and the conditions under which it was composted. Pasteurizing removes the constraints. Pasteurization can be accomplished by way of composting or through the application of an external source of heat. Except through contamination by contact (e.g., adhering compost particles), direct transfer of pathogenic organisms between members of the food chain, even without disinfection, is either nonexistent or very minor.

Types and concentrations of pathogens that *might* be in the product prior to pasteurization depend upon the feedstock. Yard wastes are not likely to contain human pathogens because human body wastes are not involved. However, they may contain organisms which are pathogenic to pets or to plants. MSW may contain some human pathogens because of contamination by body wastes. Because the indicator microorganisms for the presence of body wastes of man and animal origin are alike,[15,16] the extent, if any, of contamination with human body waste cannot be measured by concentration of "indicator organisms." In fact, the concentration of the indicators in yard waste from areas that have a sizeable pet population may rival that in sewage sludge.

Improperly composted food wastes could contain zoonotic organisms (trichina, ascaris, taenia) by way of meat scraps. In summary, composted yard waste is not likely to contain human pathogens, whereas inadequately composted food waste or sewage sludge could.

The conclusion to be drawn from the information in the preceding paragraphs is that compost products, especially those from MSW and sewage sludge, should be routinely analyzed as a precautionary measure.

## Market Development

Results of marketing studies of composts produced from municipal wastes have shown that the development of a market is to a large extent a matter of overcoming inertia and bias and instilling an awareness in potential users. This can be done through a program of education and promotion. The task is made easier by the fact that the product has a genuine utility. All that remains is for the utility to be recognized by the potential market.

Five key steps to compost marketing are (1) market analysis, (2) product quality, (3) user education, (4) promotion and demonstration, and (5) sustaining a market.

### *Market Analysis*

Market development should begin with basic market research. Once the benefits of compost are identified, a survey should be conducted to evaluate potential uses and markets for the product. The survey should be designed to collect data on what types and quantities of soil amendments are currently being used and for what purpose. Next, the survey should assess the potential interest of the respondent in using compost produced from MSW or yard waste. Data should also be collected on the specific requirements of potential users, e.g., nutrient content, particle size, pH, contamination limits, delivery requirements.

The analysis is best conducted by way of interviews and/or questionnaires to determine the needs and interest of the prospective customers. It should be remembered that knowledge of the customer's values and motivations is a necessary condition in the sale of a product.

Interviews should be conducted with representatives of groups that are a potential market for the product. Among these groups are the major agricultural sectors, government agencies, landscapers, soil vendors and distributors, nurseries, and the general public. When conducting the desired number of personal interviews becomes infeasible and if it is appropriate to do so, personal interviews can be supplemented by mailed questionnaires.

Questions should be adapted to the targeted market. Thus, information to be collected from the agricultural sector should concern types of crops grown, amounts of NPK used, times of the year when nutrients are applied, type of soil, and amount of organic matter in the soil. Other information would pertain to size and ownership of the land.

Once the assessment of the full size of the market potential is completed, a marketing plan can be developed. The marketing plan should specify the steps to be followed and should elaborate on the best method of carrying out the steps.

### *Product Quality*

As discussed earlier, product quality is a key factor in the marketability of a compost product. The decision regarding what quality of product to produce should be based upon the results of the market analysis. For example, if the primary potential markets for the compost are container nurseries and residential gardening, a high-quality compost will need to be produced. On the other hand, if the compost is intended for use as landfill cover or land reclamation, a lower quality product will likely be acceptable.

The types of feedstocks accepted at the facility will affect the quality of the compost, as will the characteristics of the feedstocks. Feedstock requirements (e.g., types and allowable percentages of contaminants) need to be developed based on intended end use of the product.

The compost process also affects the characteristics of the product and should be designed based on the specifications required by the targeted market or markets. Design features that affect product quality include shredding, screening, turning, and monitoring. In addition, the regular conduct of laboratory analyses may be needed for certain markets.

If the intent is to target markets that require a high-quality product, the processor needs to ensure by choice of feedstocks and process design that a high-quality product is produced consistently.

### *Pricing Policy*

For obvious reasons, pricing policy has a significant impact upon the development and maintenance of a market. An excessively high price places the product beyond the financial means of the user. Moreover, compost must be competitive with other organic products (e.g., mulches, mushroom compost, manures) for a share of the organic fertilizer market. If the spread between the price of compost and the competing products is too wide, the tendency of users will be to buy the less expensive product, even though they may be well aware of the benefits to be gained by using compost.

Regarding the agriculture market, the reality that agriculture is a marginally financial activity must underlie all other considerations when arriving at a selling price. Furthermore, the smaller the operation, the slimmer the margin.

A number of approaches are available for arriving at a price. One approach is to base the price on the cost to produce, distribute, and market the compost. Another approach is to follow the pricing scheme set by manufacturers of organic fertilizers or other organic soil amendments. Compost produced specifically for higher levels of use (e.g., landscaping and cultivation of ornamentals) should be priced according to the buyer's ability and willingness to pay as well as the price of competitive products.

### *Education*

An important part of market development is the establishment of a receptive "climate" by convincing the general public of the utility of using compost and by overcoming bias generally associated with waste materials.

Instilling an awareness in the public should be the first step in an area in which a market for compost either is nonexistent or is very small. Educating the public is an excellent mechanism for accomplishing that step. The educating can be done by way of presentations in the media (publications, radio, and television). Themes of the presentations may include the advantages and disadvantages of compost utilization, methods of producing compost, information on obtaining compost, and the proper utilization of compost.

Education of the general public should be followed (or accompanied) by the education of the potential users of the product. For example, this can be done by addressing farm and urban groups. These presentations should include giving assurances when needed and demonstrations regarding the best use of the product for particular applications.

As a part of the implementation of the education and promotion programs, efforts should be made to enlist the services of government and education specialists who deal with market segments on a local scale. Because of their close association with the groups, their advice is more likely to be heeded.

User education should include the provision of information regarding the usefulness and use of the product for the particular purpose of the user, e.g., health effects, plant growth and yield, and safety for food crops. For commercial markets, specifications of the compost product need to be provided as well.

### Promotion and Demonstration

Compost marketing specialists emphasize the importance of actively marketing the product through promotional activities and actual demonstrations of the use of the product. Promotion can include activities such as lectures at civic groups, displays at fairs, creative packaging, brochures, advertisements, and articles in technical journals. Demonstrating the effectiveness of compost as a soil amendment through growth tests can be particularly effective and necessary in targeting commercial markets, since these markets have a financial stake in using a new product. Growth tests can be conducted on public lands, at the processor's facility, or at actual commercial operations that volunteer to participate in such a program.

One entrepreneur states that 90% of his company's compost marketing activities are concerned with making potential users aware of the existence of his company.[17] The promotional efforts include lectures before garden clubs and presentations from a nearby radio station. The company also has purchased radio time and local newspaper space. Additional helpful steps were the acquisition of a suitable brand name and logo, and the development of a rapport with the local agriculture agent. The company's compost site is surrounded by an 8-acre market garden that attracts favorable publicity for the product. In promotional efforts, care is taken to emphasize that composting is a recycling and energy-conserving activity and, as such, transforms local liabilities into an asset.

An example of a demonstration program is the 2-year project conducted by the U.S. Bureau of Solid Waste Management at Johnson City, TN.[18] The demonstration program had two principal objectives: (1) to investigate under field conditions the potential uses of compost and the effect of each use on crop production; and (2) to introduce the product to potential users and observe their reaction to the prospect of continuing to use the compost after the completion of the demonstration program. Volunteer farmers were given sufficient compost to meet the rates specified for their respective tests. An indication of the extent of the participation is the fact that the tests involving the effects of compost on tobacco yield involved 25 farms.

Application rates investigated depended upon the crop involved in the particular test. Thus, the range of loadings was 12 to 64 tons/acre. Crops investigated in the demonstration were tobacco, corn, tomatoes, and ornamentals. Although the responses from all the farmers in terms of yields were favorable, the most favorable came from those who raised high-value cash crops.

### Sustaining a Market

Sustaining a market requires a concerted effort on the part of the processor or marketing agency. Education and promotional activities must be continued and the compost product must continue to be priced competitively. Two important factors to sustaining a compost market are product consistency and sustained availability.

*Product Quality.* Efficient crop production depends upon the use of a soil amendment of known composition and physical characteristics. Variation in consistency detracts from the utility of the product and as a consequence leads to a loss of consumer interest. Therefore, it is extremely important that the compost meet a fixed set of specifications.

*Sustained Availability.* Availability pertains to two aspects: (1) enough compost must be produced to allow it to be placed on the market, where it can be seen and tried by prospective users, and (2) it must be consistently available. Sporadic availability would be destructive to marketing efforts. The need for a sufficiently large amount of compost makes demonstration an important step in market development. The importance of constant availability is obvious. As with consistency of quality, efficient business management depends upon an assured supply of raw materials. If compost is only sporadically available, users will rely upon a competing material and lose interest in compost.

## Distribution

Distribution derives its importance from its bearing on the availability of the product to potential users. Approaches currently used or under consideration for the distribution of compost range from transportation in bulk form free of charge to the end user, to bagging and distribution through channels already established for other soil amendments.

The cost of shipping the product is a key factor in the financial feasibility of a compost operation. The modest monetary value of the product and its low bulk density combine to exacerbate the cost of long-distance transport.

Regulated freight rates typically are based upon the cost and value of service. Among the factors that can affect cost are shipping weight, liability to damage, propensity for combustion or explosion, ease of loading and unloading, and frequency and regularity of shipment. Having taken all of these items into consideration, the carrier then evaluates the value of service. In effect, the demand for the transportation service is assessed and is priced accordingly.

Transportation of secondary materials introduces additional complexities into the distribution process because of their low value. Indeed, in some cases the shipping costs may exceed the value of the material being shipped. With respect to MSW compost, shipping classifications stemming from a definition of the nature and composition of the material (e.g., compost being classified as soil) can confuse the rate establishment issue.

In general, compost can be transported by truck, rail, ship, or barge.

### *Motor Freight (Truck)*

For transport by truck, a commodity is classified according to the National Motor Freight Classification (Classification Description). Among the considerations that enter into the classification are these three: (1) the rule of analogy applies if the exact item is not referenced in the National Classifications; (2)

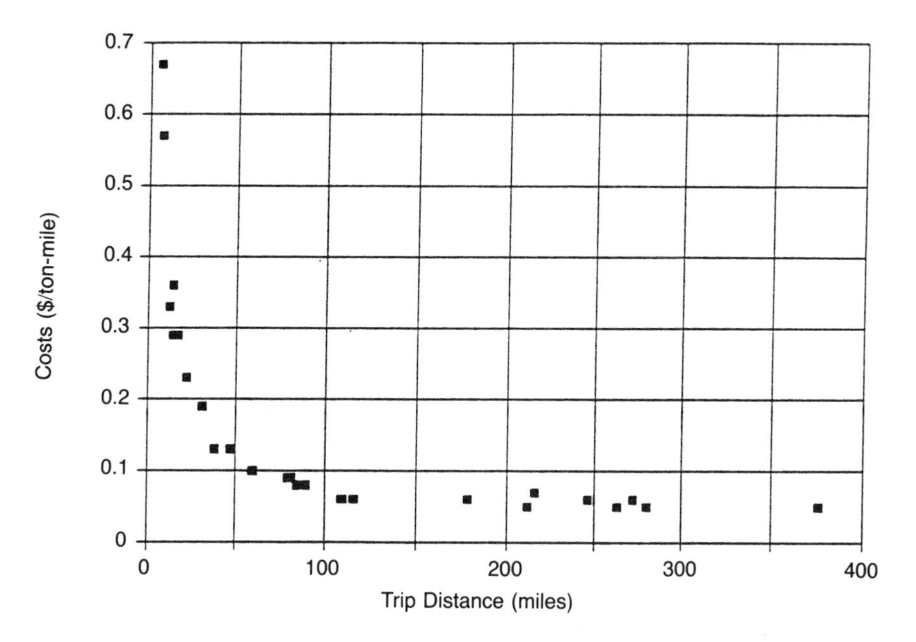


**Figure 8.1.** Intrastate motor carrier rates for bagged compost. (*Source:* Washington Department of Ecology.[5])

special rates are available because of the quantities and characteristics of recycled materials; (3) volume incentive service, special volume incentive, and premium volume may be applicable as possible special rate classifications.

According to the National Motor Freight Classifications, compost would be classified as soil and would be considered in Class 50 LTL (less than truckload) and Class 35 TL (truck load).

To qualify for special rates, the total load should be ready to load at one time. The particular classification depends primarily upon the total weight of the shipment, loading and unloading restrictions, and the limits of liability.

An example on the variation of motor freight rates as a function of distance is presented in Figure 8.1.

## *Railroad Freight*

For shipment via railroad freight, commodities are classified according to the Unified Freight Classification-12 (UFC-12). If the exact item is not referenced in UFC-12, the rule of analogy applies (e.g., compost classified as soil or agricultural mulch).

As with motor freight rates, railroad rates are governed by a number of factors. For instance, the rates for compost transported within the state of California are a function of the minimum weight shipment per railroad car. The minimum weight for compost is on the order of 100,000 lb.

### Ocean Shipping

Pacific Westbound Conference is a cartel of various member shipping lines, and rates are standardized for all of its members. Basically, two rates are available: (1) contract rates are established for a certain commodity under the agreement that all future shipments of that commodity will be shipped by the chosen shipper; (2) noncontract rates do not carry the restriction on future shipments by the same shipper. In general, contract rates run 5 to 10% lower than noncontract rates. Conference tariffs form the working basis for commodities shipped from the West Coast to the Far East.

Independent shippers use the Conference rates as the upper limit and, depending upon the commodity, generally offer lower rates. A rule of thumb of the shipping lines purports independent rates to be 10 to 15% below Conference rates.

These rates reflect shipment in 20- or 40-ft containers. Careful coordination between consignor and shipper is necessary to achieve the maximum amount of benefit from containerized cargo.

### Transport by Barge

Hauling compost can be done by barge in certain regions of the U.S. Judging from information obtained from a West Coast barging company, barging rates generally seem to be independent of the commodity being hauled and are primarily a function of the weights transported and the crew time involved.

Some pertinent distances and running times for a shallow draft tug of 1000 hp are summarized as follows:

| Round Trip Distance (Nautical Mi) | Running Time (hr) |
|---|---|
| 92 | 16 |
| 100 | 18 |
| 124 | 22 |
| 146 | 30 |

If 4000 tons/day of compost were to be transported by a barge having dimensions of 250 x 72 ft with steel bins of 230 x 50 x 20 ft, a total capacity of 8520 cu yd would be needed. Using a density of 500 lb/cu yd, the bin would contain 2130 tons, with 6 to 7 ft draft on the barge. The cost of a new barge of this size and configuration is about $2.75 million based on 1992 prices.

The following considerations entered into the estimation of the cost per ton of barging service:

1. The barges are ''bare-boat'' charter at a rate of $1600 per day. Insurance and maintenance are the responsibility of the charterer.
2. Based on current labor and fuel costs, the daily rate for a 1600-hp tug for this operation would be about $7000. This rate includes fuel.
3. The estimated cost for one tug and four barges hauling 60,000 tons/month of material is about $6.70 per ton.

## RECYCLABLES

In general, marketing of recycled materials can be accomplished through two means: (1) direct sale to manufacturers who will use the material in their process; and (2) sale to intermediaries (dealers/processors, wholesalers, brokers) who may process the material further or simply broker it to the manufacturers. Within each of the methods of marketing, there is a wide range of potential markets available for a given recyclable, depending largely upon the type and quality of the recycled material, and the location of the recyclables.

Manufacturers that use recycled materials as feedstocks (either with virgin materials or in place of them) do so because of the resulting benefits. Examples of cost benefits include reduced capital requirements, reduced energy requirements, reduced raw material needs, and more accessible or less costly sources of supply.[19] In addition to cost benefits, environmental and/or promotional benefits may prompt manufacturers to utilize recyclables in their operations.

Table 8.12 presents a listing of industries that use recycled materials as feedstocks, the virgin materials that are replaced, and the benefits realized by using recycled feedstocks.

The following sections discuss the potential uses for some of the major components of the waste stream. Some of the discussions include brief and simplified descriptions of the conventional processes involved in the manufacture of the particular commodity. These discussions are included for completeness, to satisfy the curiosity of the reader, and to provide the reader with at least an understanding of the terminology. In addition, the presentation provides an indication of some of the chemicals used in the manufacture of a particular commodity and how these chemicals can have an impact on a recycling process.

### Potential Uses

#### *Paper*

Cellulose is the most abundant organic substance available in nature. Cellulose is also one of the most versatile, simple, and renewable materials known. The material is converted to paper products by the pulp and paper industry. In the conversion process, hundreds of tons of chemicals are used to produce one ton of paper.

Writing paper was first manufactured between 2500 and 3000 B.C. by using a reed that grew along the Nile called *papyrus,* and thus the word paper. The Chinese invented the actual manufacture of paper around 105 A.D. However, the process did not undergo major improvements and did not become known to southern Europeans until the end of the 14th century. The first paper mill was established in North America in 1690. Since then, the pulp industry has evolved into three methods: mechanical, sulfite, and soda.[22]

The production of finished paper involves two major phases: (1) the manufacture of various pulps; and (2) the conversion of the pulp into paper. The major fraction of the pulping is conducted by chemical means. In the process, chemicals

**TABLE 8.12 Examples of Industries Using Recycled Materials as Feedstocks**

| | Type of Industry | Virgin Materials | No. of Manufacturers | Consumption (1989) (million tons) | Benefits |
|---|---|---|---|---|---|
| Paper | Paper and paperboard mills | Wood pulp, other plant fibers | 700 | 85 | Less costly feedstock<br>Reduced water consumption<br>Lower capital requirement |
| Glass | Container manufacturers | Sand, limestone, soda ash | 16 | 11 | Energy savings<br>Cleaner furnace firing<br>Potential furnace life extension |
| Tin cans | Steel mills and detinning plants | Iron ore, lime, coke | 50–150 | 80 | Reduced capital requirements<br>Reduced raw material requirements<br>Higher quality input (if purchased from detinners) |
| Aluminum cans | Primary and secondary aluminum producers | Refined bauxite, carbon | 60+ | 8 | Avoided capital cost<br>Reduced energy consumption |
| Plastic | Plastic resin, film, and fiber manufacturers and processors | Products of petroleum and natural gas distillation and refinement | 14,000+ | 30 | Cost savings<br>Availability |

*Source:* Finelli.[19]

are used to dissolve the lignin from the cellulose fibers. The main source of cellulose for the manufacture of paper is wood. Other materials also are used for paper making; some of these materials include: cotton, linen rags, bagasse, various fibers, and used (secondary) papers.

Groundwood is produced without any chemical treatment. In the process, soft wood is first slashed and debarked. The wood is then ground in water and put through a series of flotation and screening processes. The fines that pass through screens are transported to thickeners which provide the commercial mechanical pulp. The major energy consumption in this process is primarily due to size reduction. Traditionally, groundwood has been used to manufacture lower grades of paper and board, particularly when longevity is not required. Paper made from groundwood deteriorates relatively fast due to the chemical decomposition of the noncellulosic fractions of the wood. In some cases the quality of mechanical pulp is increased by mixing with small amounts of chemical pulp. Groundwood is bleached to achieve desired degrees of brightness. The material is bleached by using bleaching agents such as ozone, chlorine dioxide, hydrogen peroxide, and sodium peroxide.

Kraft or sulfate pulp is the method used to produce a major fraction of the pulp. In this process, almost any type of wood can be used. The process originally was developed to remove the large concentrations of resins and oils in the woods. The kraft process involves slashing and debarking of the logs. The logs are then chipped by large, rotating disks, screened, and introduced into a digester. The digester is filled with essentially sodium sulfide and caustic soda. Live steam is injected into the digester and the pressure is raised to about 110 psi. The "cooking" period lasts for about 3 hr. After cooking, the contents of the digester (brown stock) are discharged. The pulp is separated from the cooking liquor and washed. The cooking liquor (black liquor) is processed to recover the chemicals in it. The pulp undergoes a series of processes aimed at removing impurities. The clean pulp is bleached, neutralized, washed, and thickened to prepare it for the manufacture of sheets. Although the various types of pulps frequently are manufactured into coarse sheets, they lack the properties desirable of finished paper. These properties include: strength, opacity, proper surface, and feel. Pulp is prepared for formation into paper by two general operations: beating and refining. Filler, sizing, and coloring are then added. A filler is added to occupy the spaces between the fibers and thus give the paper a smoother surface. Fillers are finely ground inorganic substances such as talc, some clays, or titanium dioxide. Sizing is added to the paper to give it resistance to penetration by liquids. Of course, some papers such as blotters would not need the addition of sizing. Sizing consists of either a soap or a wax followed by the precipitation of the size with paper maker's alum. Coloring is added to achieve the desired color of paper. The sheets of paper are formed from slurries containing less than 1% fiber stock in Fourdrinier machines.

Paper and paperboard products are not consumed in the strict sense of the term; instead they are used and discarded and therefore acquire the designation "wastepaper." Consequently, wastepaper is transformed rather than manufactured.

Wastepaper is generated as an indirect result of a manufacturing or utilization process and must be disposed of in some manner. Generally the time scale for disposal is short, on the order of days. Although suppliers are faced with problems of this nature, consumers, on the other hand, would like to order and receive paper stock in a quantity and manner in accordance with their needs. This basic situation is manifested in the cyclical nature of the supply and demand, which constitutes a serious problem in the paper market. Quality is yet another point of conflict between the supplier and the consumer.

As previously indicated, used paper can be processed to make new paper. The total consumption of wastepaper (all grades) in the U.S. has increased from about 12 million tons in 1970 to approximately 22 million tons in 1990. About half of the wastepaper consumption is attributed to cardboard.[21] The various types of paper that can be recycled include newsprint, corrugated containers, magazines, high-grade, and mixed paper. There are also a number of potential uses, depending on the material type and quality.

The paper stock industry is the principal vehicle for the recycling of paper. Used paper is moved through the collecting, sorting, grading, packing, and shipping processes which are within the overall organization of the paper stock industry. In order to insure some type of order in marketing, the Paper Stock Institute of America publishes a set of specifications for 49 grades and 31 specialty grades of wastepaper (Circular PS-86).

Marketing patterns for secondary fibers have evolved from a rubric of small, independent dealer operations servicing a few customers to nationwide operations. Several of the large paper stock companies are subsidiaries of paper mill companies and, as such, serve as the source of secondary fiber for the mill. These types of paper stock companies generally deal only with specific grades. There are several factors that contribute to the erratic demand for paper stock. Some of these factors include:

1. Difficulty for mills to carry extensive paper stock inventories.
2. Paper stock demand fluctuates as demand for paper, paper board, and/or converted products fluctuates. Periodic mill shutdowns can lead to seasonal variations in demand.
3. Paper stock is considered a perishable item and can be hazardous to handle and store.
4. Space for storage of paper stock is both limited and costly.
5. Any time the demand for paper stock is low, some collectors go out of business.

These factors have an impact on marketing practices. Marketing of wastepaper in the U.S. also is impacted by the quantity of materials exported. The quantity of wastepaper exported from the U.S. has increased from about 408,000 tons in 1970 to more than 6.5 million tons in 1990. Of the quantity exported in 1990, about 62%, or more than 4 million tons, are exported to the Far East and Oceanic. Typically, corrugated is the type of wastepaper that is primarily exported to those areas. The regions of the U.S. that export the most wastepaper are the Mountain, the Pacific, and the Mid-Atlantic.[21]

Domestic demand for recycled paper is created largely by four major industries: paper products, paperboard products, construction paper, and molded pulp products. Listed below are examples of materials produced by these industries:

- *Paper products* — newsprint, printing and writing papers, packaging, bags and sacks, tissue and toweling
- *Paperboard products* — facing material for corrugated boxes, corrugating medium, shoe boxes, file folders
- *Construction paper* — roofing materials, acoustical tile
- *Molded pulp products* — egg cartons, fruit packing layers

Other, but generally smaller, markets for recycled paper include cellulose insulation, animal bedding, hydromulching, and the production of liquid and/or solid fuels.

The lack of economical production capacity recently has become a barrier to market expansion for newsprint. However, advances in deinking technology may make it more economically feasible than before to produce newsprint using a high percentage of recycled news as feedstock.

A summary of the total wastepaper utilization and recovery in the U.S. is presented in Table 8.13. As shown in the table, the total recovery rate has increased from about 22% in 1970 to more than 33% in 1990. The data in the table also show that consumption at domestic paper and paperboard mills accounts for more than 21 million tons of wastepaper usage.

### *Glass*

Glass has a variety of applications because of its physical and chemical properties. Glass is transparent, is an effective electrical insulator, has high resistance to chemical attack, has the ability to contain a vacuum, is impermeable to most liquids and gases, and can be molded to nearly any shape and size.

The discovery of glass is very uncertain. It has been estimated that the Egyptians were making jewels of glass as early as 6000 or 5000 B.C. Window glass is mentioned at about 290 A.D. However, the use of window glass did not become common until the 16th century. Glass works in the U.S. were founded in Jamestown, Virginia in 1608. Prior to 1900 the manufacture of glass was an art. Formulas were closely guarded and processes for manufacture were based largely on experience.[22] Based on approaches patented in the U.S. in 1902 and 1905, scientists in England perfected the float glass process of manufacturing glass. Shortly thereafter, the float glass process took over the window glass market. Since then, major advances have been made in the manufacture of new products and in the use of automation for the production of containers and other materials.

A classic definition of glass is as a rigid, undercooled liquid having no definite melting point and a high viscosity which prevents crystallization. The major ingredients in the manufacture of glass are sand, lime, and soda ash.

TABLE 8.13 U.S. Waste Paper Utilization and Recovery - Total All Grades (Thousands of Short Tons)

| | Paper & Paperboard | | Waste Paper Usage | | | | | | |
|---|---|---|---|---|---|---|---|---|---|
| | | | Domestic Consumption | | | | | Total Recovery | |
| Year | Production[a] | New Supply[b] | At Paper & Paperboard Mills | Utilization Rate (%)[c] | All Other Uses[d] | Exports | Imports | Total Recovered[e] | Recovery Rate (%)[f] |
| 1970 | 51,670.1 | 55,967.9 | 11,803.0 | 22.8 | 418.0 | 408.2 | 67.0 | 12,562.2 | 22.4 |
| 1971 | 53,162.0 | 57,449.9 | 12,106.0 | 22.8 | 442.0 | 419.0 | 68.0 | 12,899.0 | 22.5 |
| 1972 | 57,433.7 | 62,040.7 | 12,925.0 | 22.5 | 447.0 | 415.0 | 88.0 | 13,699.0 | 22.1 |
| 1973 | 59,900.0 | 65,003.9 | 14,094.0 | 23.5 | 499.0 | 683.4 | 87.0 | 15,189.4 | 23.4 |
| 1974 | 59,040.2 | 63,308.2 | 13,982.0 | 23.7 | 489.0 | 1,307.1 | 89.0 | 15,689.1 | 24.8 |
| 1975 | 50,976.1 | 54,113.0 | 11,748.0 | 23.0 | 535.0 | 861.4 | 72.0 | 13,072.4 | 24.2 |
| 1976 | 58,328.7 | 62,013.9 | 13,622.0 | 23.4 | 630.0 | 1,272.7 | 106.0 | 15,418.7 | 24.9 |
| 1977 | 60,040.1 | 64,242.8 | 14,058.0 | 23.4 | 870.0 | 1,512.4 | 92.0 | 16,348.4 | 25.4 |
| 1978 | 62,046.7 | 67,787.1 | 14,760.0 | 23.8 | 502.0 | 1,612.9 | 70.0 | 16,804.9 | 24.8 |
| 1979 | 64,345.0 | 69,796.0 | 15,361.0 | 23.9 | 509.0 | 2,126.6 | 78.0 | 17,918.6 | 25.7 |
| 1980 | 63,600.0 | 67,166.2 | 14,922.0 | 23.5 | 472.0 | 2,636.2 | 87.0 | 17,943.2 | 26.7 |
| 1981 | 64,258.6 | 67,956.7 | 15,037.0 | 23.4 | 480.0 | 2,282.0 | 79.3 | 17,719.7 | 26.1 |
| 1982R[g] | 60,951.7 | 64,729.7 | 14,433.0 | 23.7 | 487.0 | 2,232.6 | 74.3 | 17,078.3 | 26.4 |
| 1983R | 66,748.7 | 71,166.2 | 15,648.0 | 23.4 | 474.0 | 2,704.9 | 100.4 | 18,726.5 | 26.3 |
| 1984R | 70,248.5 | 76,936.6 | 16,723.8 | 23.8 | 459.2 | 3,456.3 | 109.7 | 20,529.6 | 26.7 |
| 1985R | 68,688.2 | 76,138.4 | 16,371.0 | 23.8 | 529.1 | 3,555.9 | 87.5 | 20,368.5 | 26.8 |
| 1986R | 72,505.4 | 79,755.0 | 17,934.2 | 24.7 | 593.5 | 4,092.5 | 99.3 | 22,520.9 | 28.2 |
| 1987R | 75,958.5 | 83,491.3 | 18,693.8 | 24.6 | 657.2 | 4,809.2 | 127.4 | 24,032.8 | 28.8 |
| 1988R | 78,084.4 | 85,503.4 | 19,684.5 | 25.2 | 703.0 | 5,952.7 | 161.4 | 26,178.8 | 30.6 |
| 1989R | 78,355.3 | 85,155.5 | 20,220.2 | 25.8 | 722.0 | 6,307.0 | 172.6 | 27,076.6 | 31.8 |
| 1990P | 80,402.4 | 86,756.5 | 21,791.5 | 27.1 | 753.0 | 6,504.9 | 122.5 | 28,926.9 | 33.3 |

*Source:* American Paper Institute.[21]

[a] Paper and Paperboard Production includes all grades of paper, paperboard, wet machine board, construction paper and board, excluding hard pressed board.
[b] New Supply equals production plus imports less exports, excluding hard pressed board (imports and exports include paper and paperboard converted products).
[c] Utilization Rate is the ratio of waste paper consumption to total production of paper and paperboard.
[d] All Other Uses includes molded pulp and estimated consumption of newsprint grades for highway seeding, animal bedding, cellulose insulation, shredding packaging protection and other end uses.
[e] Total Waste Paper Recovered is the sum of consumption at paper and paperboard mills, other uses and exports less imports.
[f] Recovery Rate is the ratio of total waste paper recovered to new supply of paper and paperboard.
[g] R: revised; P: preliminary.

Several types of glasses are manufactured. Some of these glasses include soda-lime, lead glass, fiber glass, safety glass, and others. These glasses are manu-manufactured by using large quantities of glass sand. In order to flux the silica, soda ash, salt cake, and limestone or lime are necessary. The process also requires several other chemicals such as metallic oxides and salts. The sand used for glass manufacture must be almost pure quartz. Due to the relatively large quantities of sand required, generally the location of the glass-sand deposit dictates the location of the glass factory. Since iron has a negative impact on the color of glass, the concentration of iron in the sand should be less than 0.5%. The manufacture of glass can be divided into four main stages: melting, shaping or forming, annealing, and finishing. In general, the raw materials are introduced into a furnace and melted at temperatures on the order of 2600°F. Once melted, the glass is shaped by either machine or hand molding. The particular item must be shaped in a very short time since during this time the glass changes from a relatively viscous liquid to a clear solid. It is then apparent that the design of the machines for shaping glass requires attention to a number of complicated factors such as heat transfer, clearance of bearings, stability of metals, etc. In order to reduce strain, it is necessary to anneal all glass items. Annealing involves two steps: maintaining the glass above a certain temperature for a minimum period of time to reduce internal stress, and cooling the glass to ambient temperature at a certain rate such that the strain is maintained below a certain maximum level. Finally, the glass items must be finished. Finishing involves a series of steps which include grinding, polishing, and cutting.

Glass manufacturing has traditionally been a fully integrated process. The integration begins with the mining of the raw material and ends with the finished product at the same general location.

Currently, the primary market for glass is in the manufacture of container glass. Glass used in the manufacture of light bulbs, drinking glasses, cookware, and flat glass (mirrored and window glass) is of a different composition than container glass, and as such, these products are considered contaminants to the container glass industry. Color-sorted glass (crushed and free of contaminants) is required in glass container manufacturing. Color is important because it has an impact on the light transmission through a container and thus the relative stability or durability of the contents. In addition, large quantities of contaminants in a particular batch can negatively impact the melting furnace. Contaminants also impact the integrity of the container.

The municipal waste stream contains on the order of 10% glass (by weight). This concentration includes the materials that eventually are returned or recycled. The actual concentration of glass in the waste stream disposed at a landfill varies substantially from community to community. Containers make up about 10% of the concentration of glass in the waste. It has been estimated that cullet could comprise as much as 30% of the raw material introduced into a melt for the manufacture of glass containers. The specifications or standards are dictated by the type of material that is going to be manufactured. In the manufacture of containers, the standards are quite stringent and well defined. Nearly every type of glass manufacturing process uses a certain quantity of cullet generated internally (primary cullet). The quantities of primary cullet vary from 8 to 100%.

The manufacture of glass containers traditionally uses on the order of 15%. At present, the use of cullet is estimated at 25 to 30%. The industry has announced an overall goal of 50% and has established several recycling programs.

There are several advantages in the use of cullet. Cullet liquifies at a lower temperature than the raw materials used in the manufacture of glass. This results in energy savings, reduction of air pollution, extension of the life of the furnace linings, and the production of a faster melt. The specifications for the use of secondary glass as cullet for the manufacture of glass containers are presented below.

1. **Color sorting:** Permissible color mix levels are

| | | |
|---|---|---|
| Flint glass | 95–100% | Flint |
| | 0–3% | Amber |
| | 0–1% | Green |
| | 0–1% | Other |
| Amber glass | 90–100% | Amber |
| | 0–10% | Green |
| | 0–5% | Flint |
| | 0–5% | Other |
| Green glass | 80–100% | Green |
| | 0–15% | Amber |
| | 0–10% | Flint |

2. **Ferrous (magnetic metal):** Loads will be rejected if there are
   A. Any pieces larger than 6 x 6 x 12 in.
   B. More than 1% of the load is smaller than 6 x 6 x 12 in., and larger than 1/2-in. pieces
   C. More than 0.05% of the load is smaller than 1/2-in. pieces

3. **Nonferrous metal (aluminum, lead, etc.):** Loads will be rejected if there are
   A. +3/4-in. glass packaging material (closures, aluminum foil) greater than normal amounts inherent to glass packaging
   B. +3/4-in. nonglass packaging material (lead, copper, brass) greater than 0.5%

4. **Organic material (labels etc.):** Loads will be rejected if there are
   A. Glass packaging materials (labels, Plasti-Shield) greater than normal amounts inherent to glass packaging
   B. Nonglass packaging materials (paper, wood, rubber) greater than 0.5%

5. **Refractory material (ceramics, pottery, tableware, tile, etc.):** Loads will be rejected if there are
   A. Any particles in a 50-lb sample larger than U.S. 8 mesh
   B. More than 1 particle in a 50-lb sample smaller than U.S. 8 mesh, but larger than 20 mesh U.S.
   C. More than 40 particles in a 50-lb sample smaller than U.S. 20 mesh but larger than 40 mesh U.S.

6. **Cullet sizing:** Loads will be rejected if
   A. More than 25% of the cullet is smaller than 3/4 in.

7. **Other contamination:** Loads will be rejected for
   A. Excessive amounts of dirt, gravel, asphalt, concrete, limestone, garbage, etc.
   B. Excessive amounts of moisture
   C. Contamination caused by burning glass containers
   D. Pyrex, oven-ready material, plate glass, automobile glass, light bulbs, etc.

Secondary glass can also be used by several other industries such as glass bead manufacturing, roadway materials, building materials (fiberglass insulation), and for specialty glass. These industries generally accept mixed-color cullet.

### *Ferrous Metals*

Steel traditionally has been manufactured using one of the following methods: blast furnace (in which pig iron is produced from ore, coke, and fluxes), the open-hearth furnace, the basic oxygen furnace, the electric furnace, and the cupola furnace (used for the production of cast iron).

In general, there are three types of iron and steel scrap. They are home scrap, prompt industrial scrap, and obsolete scrap. Home scrap is that generated internally by the iron and steel industry during production. Home scrap is recycled internally and never reaches the scrap market.

The concentration of ferrous metals in the municipal waste stream is on the order of 3 to 7%. One of the major components of ferrous wastes is the steel can. There are basically two types of steel cans in the waste stream. The food container or "tin can" and the bimetallic can (steel can with an aluminum top). Food and beverage containers account for the major fraction of the tin cans discarded.

*"Tin" Cans.* "Tin cans" is the popularly used term for food, beverage, and other common household, commercial, and institutional containers whose base metal is steel. The steel cans are coated with tin to minimize corrosion, stabilize flavors, and ensure the quality of their contents.

Tin cans are usually marketed directly to steel mills or detinners, or through intermediate processors. Detinners remove the tin plating and then sell the material as a higher grade material to steel mills. Current steel-making techniques can accommodate 5 to 10% tin in the infeed without making the product too brittle.[20]

The specifications for purchasing postconsumer steel cans vary from market to market. The following is a listing of general specifications for the forms of steel cans normally purchased.

Can scrap may be baled; the bales should be 2 x 2 x 2 ft (or 3 ft) in size. The density of each bale should be between 75 and 80 lb/cu ft. Bale integrity must be maintained during shipping and magnetic handling at the mill. The cans may be baled without removing the paper labels. However, the bales must be free of nonmetallic materials such as plastics, water, dirt, wood, and debris.

Densified (biscuit) can scrap for steel companies should be stacked and banded into bundles. The bundles should have a density of between 75 and 80 lb/cu ft. Bundle weight is subject to negotiation.

Baled can scrap for detinning companies may be of varied dimensions. Density should be nominally 30 lb/cu ft. Higher densities are subject to negotiation. Wire or steel banding is acceptable.

Loose cans (whole or flattened) as well as shredded cans (loose or baled) are acceptable, subject to negotiation.

The Steel Can Recycling Institute advises that it is important to contact regional buyers before processing the cans in order to ensure that the preparation process will be acceptable. More detailed information on specifications for steel can scrap may be obtained by reviewing the specifications prepared by the American Society for Testing and Materials (E-702-79, E701-80, and E-1134-86).

*Other Ferrous.* This category of recycled materials includes scrap from discarded products and demolished structures. Examples of other ferrous are white goods and automobiles. The recycled material is usually shredded and marketed directly to steel mills or through salvagers.

### *Aluminum*

Aluminum makes up about 8% of the solid portion of the crust of the earth and is probably one of the most abundant metals in the world. Nearly every country has large supplies of materials that contain aluminum. However, processes for obtaining metallic aluminum from these compounds are not as yet economical. Aluminum metal was first isolated in metallic form in 1825. In 1886 the first aluminum was produced by the present large-scale process. The aluminum industry has grown steadily based primarily on new and expanding markets established by the research and development efforts of the industry.

Metallic aluminum is produced by the electrolytic reduction of pure alumina in a bath of fused cryolite. Electrolysis takes place in very large, lined containers called cells. Inside the cells are the cathode compartments. The solution of alumina in molten cryolite is electrolized to form metallic aluminum. It serves as the cathode. The molten aluminum is tapped from the cells, alloyed (if needed), cast into ingots, and cooled.

The utilization of the total quantity of aluminum produced each year is primarily for building, transportation, electrical, and packaging.

Aluminum scrap can be divided into two major categories: (1) new scrap; and (2) old scrap. New scrap is that generated in the production of aluminum and aluminum products. Old scrap is the material recovered from obsolete objects such as old siding, automobiles, and aluminum cans.

Virtually all new scrap is recycled. Old scrap is more dispersed than new scrap and consequently is recycled only when its market value exceeds the cost

**TABLE 8.14 Recycling Rates of Aluminum UBC in the U.S.**

| Year | Million Pounds | Billion Cans | Recycling Percent |
|---|---|---|---|
| 1972 | 53 | 1.2 | 15.4 |
| 1980 | 609 | 14.8 | 37.3 |
| 1985 | 1245 | 33.1 | 51.0 |
| 1989 | 1688 | 49.4 | 60.8 |
| 1990 | 1934 | 54.9 | 63.6 |

*Source:* U.S. EPA.[24]

of recovery. Most of the old scrap is used by smelters. The main exception is recycled aluminum cans, which are purchased by primary producers.

One of the problems associated with aluminum recycling is related to the alloying elements which become a part of the inventory of consumer items manufactured from aluminum.

*Aluminum Cans.* The use of recycled aluminum results in substantial energy savings. The major aluminum companies established recycling programs in the early 1960s. Initially, some of these programs were viewed primarily as public relations efforts. However, that view has changed. The aluminum collection and recycling programs have grown in size and number and now play a significant role in the aluminum industry. The aluminum industry can probably utilize all available scrap recovered from recycling centers and processing plants, particularly if the quality is high enough for reuse in wrought products.

The changes in the level of recycling aluminum used beverage cans (UBC) in the U.S. is presented in Table 8.14. The data in the table show that the amount of aluminum recovered has increased from about 53 million lb in 1972 to more than 1930 million lb in 1990. The quantity of aluminum UBC recovered in 1990 is equivalent to a recycling rate of about 64%.

Recycled aluminum beverage cans are generally marketed to aluminum can manufacturers where the material is processed into can sheet for the production of beverage containers. Aluminum mills constitute another market segment where recycled aluminum can be used in the manufacture of other products made from aluminum.

The demand for aluminum beverage cans and aluminum scrap is subject to the local economic climate, as well as national and international economic conditions. Demand is also seasonal, and producers often vary can sheet production when the consumption of soft drinks and beer declines.

*Other Aluminum.* Examples of materials that fall into this category are discarded pots and pans, lawn furniture, lighting fixtures, and venetian blinds. These recycled materials are generally not used for can sheet or the production of aluminum foil, but are either marketed directly to aluminum mills for the production of other products or to salvagers.

### *Plastic*

The development of plastics from the laboratory to a large number of materials tailored to the needs of industry revolutionized the construction and design engineering professions. Because of the physiochemical properties of plastics, they have displaced substantial quantities of metals, wood, and natural fibers in the manufacture of products and in packaging materials. This has led to a projected increase in the use of plastics from an estimated 6 billion lb in 1960 to more than 50 billion lb in the 1990s. Although plastic materials have been produced for more than 100 years, the industry was not well established until the early part of this century when phenolic resins were produced commercially.

At present, plastics are made from two or more of the following material groups: binder, filler, chemical intermediates for resins, plasticizers, dyes and pigments, catalysts, and lubricants. The binder generally is a resin or a cellulose derivative. The filler can be from organic or inorganic origin. Currently, much effort is being placed at developing biodegradable or thermodegradable plastics.

Although there is a wide spectrum of plastic products, there are two main types of plastic polymers: thermosets and thermoplastics. Thermosets usually are rigid due to the use of polymers which have a crosslinked molecular structure. Some thermoset plastics include: polyurethane foam and epoxy resins. On the other hand, thermoplastics are made up of single-chain polymers that soften when heated and can be reshaped. Thermoplastics are frequently used to manufacture packaging materials and account for approximately 80% of the discarded postconsumer plastic waste.

Unfortunately, the same characteristics that make plastics versatile and durable also constitute serious problems in the development, design, and implementation of suitable methods for their treatment and disposal. The concentration of postconsumer plastics in the waste stream ranges from 5 to 8%. In some areas, the concentration of plastics in the waste stream can reach 12%. About 80% of the discarded plastics is in the form of packaging, and the remaining 20% is durable plastic attributed to the increase in the use of disposable plastic materials. Recent estimates indicate that approximately 14 million tons of plastics are discarded each year.

Several types of plastics are typically recycled, including PET (2-L bottles), HDPE (milk and juice containers), film, and mixed plastics. Individual plastic resins such as PET and HDPE are of greater utility and value and can be marketed directly to product manufacturers.

Recycled PET is used in the manufacture of synthetic insulation and strapping; small molded products such as handles and audio and video cassette casings; and as an additive in the manufacture of plastic panels for insulation, of furniture, and of sporting goods. Recycled HDPE is used to produce base cups for soft drink bottles, flower pots, recycling bins, and a variety of other containers and products.

Mixed plastics often are marketed to intermediate processors where the plastics are separated into individual resins or extruded for the production of plastic products. Industries that manufacture products directly from postconsumer plastics with minimal processing include those that manufacture durable, weather-resistant items such as docks and pilings, fences, outdoor furniture, pallets, parking lot stops, and various sizes of plastic lumber. These technologies are beneficial for plastics recycling because of their ability to accept commingled resins.

### *Textiles*

Textiles are generally recycled in one of two ways. The better quality materials are often discarded to reuse centers such as thrift shops. Lesser quality textiles are processed by the rag industry into rags or fiber stock.

## Marketing Issues

A number of marketing issues should be considered when designing a facility to process recyclables or planning a recycling program. Such issues include (1) product quality, (2) product quantities, (3) market share, (4) market development, and (5) flexibility.

### *Product Quality*

Production of a high-quality product has a number of advantages. First, it allows the producer to find the highest and best use for the material. For example, the potential markets for individual plastic resins are greater than for mixed plastics. Second, high-quality products command the highest prices. Prices for uncontaminated color-sorted glass cullet are higher than mixed-color cullet or color-sorted cullet that is contaminated. Third, it increases market stability. A manufacturer is more likely to keep a supplier that provides a high-quality material.

Examples of the unit revenues for various recyclables are presented in Table 8.15. The prices are the result of a study conducted in New York City and demonstrate the wide variation in revenues that can be realized for different materials as well as different grades of material.

### *Product Quantities*

A knowledge of the types and quantities of materials available to the facility or program is critical to market development. This requires an analysis of the waste stream and an understanding of short- and long-term trends. Another important factor is whether or not there are any seasonal fluctuations in recyclables composition or quantities.

### *Market Share*

An assessment of the relative influence or position in the marketplace of the facility or program is important to know. If the facility is a large player with respect to regional demand, market development becomes a possibility and the

**TABLE 8.15 Sample Unit Revenues for Various Recyclables**

| | Revenues ($/Ton) |
|---|---|
| Paper | |
| Newsprint #6 | 0 |
| Newsprint #8 | 20 |
| Magazines/glossy | 5 |
| Glass | |
| Clear container | 45 |
| Green container | 5 |
| Brown container | 20 |
| Glassphalt | 7 |
| Plastics | |
| Clear HDPE container | 120 |
| Colored HDPE container | 60 |
| Clear PET containers | 140 |
| Green PET containers | 120 |
| Polypropylene | 120 |
| Films & bags | 0 |
| Metals | |
| Food container | 50 |
| Beverage cans | 850 |
| Bimetal cans | 50 |
| Food container/foil | 250 |
| Misc. aluminum | 400 |

*Source:* New York City Department of Transportation.[20]

processor may be able to have an influence on procurement policies. It offers both risks and opportunities. On the other hand, if the facility or program is a small player, the processor may need to target established uses and compete by offering higher product quality or lower prices.

### *Market Development*

Market development becomes possible and sometimes necessary when the supply is substantial in terms of local and regional demand. One example of recent efforts in market development is the use of shredded newsprint in place of straw for animal bedding. Studies in Iowa demonstrated not only that newsprint suppressed bacteria growth, but that lesser quantities of newsprint were needed. Efforts are also being made with regard to plastics recycling. Grocery sacks are being produced out of film plastics, and plastic lumber is being manufactured from mixed plastics.

### *Flexibility*

Recycling programs and facilities should be designed with as much flexibility as possible. Flexibility allows for different markets for the same materials as well as for changing market demands. However, flexibility has trade-offs in terms of cost.

## Marketing Strategies

The successful marketing of recyclables requires an understanding of the factors that affect the short- and long-term demand for the materials, of the requirements for specific industries, of the local and regional supply and demand, and of market development opportunities.

### Short- and Long-Term Demand

A number of factors affect the demand for the recyclables and the selling price, including (1) demand and prices paid for end products; (2) production capacity of the buying industry; (3) limits on substitutability of recyclables for virgin feedstocks; (4) availability and cost of virgin feedstocks; and (5) available supply of secondary materials. These factors are subject to change over the long term and can be affected by investments in new capacity to handle recycled feedstocks, changes in regulatory requirements, and technological developments.[19]

### Feedstock Requirements

The recovered material must meet the specifications of the intended market. Product specifications are affected by the characteristics of the incoming material (e.g., commingled, source separated) and by the methods utilized to process the material (e.g., shredding, baling). Thus, the desired quality of the product depends on the specifications of the market. Production of a higher quality product than required would likely result in increased recycling costs. Conversely, if specifications are not met, the load may be rejected outright by the buyer or subjected to price adjustment.

Specifications may be generic to the industry or they may be unique to a particular buyer. Tables 8.16 through 8.18 present examples of buyer specifications for newspaper, PET, and HDPE, which demonstrate the variation in the requirements of buyers.[20] Some industries do not allow for contamination. For example, a small quantity of impurities in glass cullet can damage a glass furnace and weaken the glass container. Similarly, the unintended mixing of plastic resins can damage the structural integrity of thermoformed plastics.[19]

### Example of a Marketing Study

A marketing study needs to be conducted primarily to assess the local supply and demand, regional supply and demand, and market development opportunities.

The results of a marketing study conducted recently for a large metropolitan area for recycled newspaper[20] indicated the following:

1. The maximum benefit in terms of long-term price stability and ability to reliably market newsprint would result from maximizing the amount of high-quality newsprint.

**TABLE 8.16 Examples of Buyer Specifications for Newspaper**

| | | | | | Contamination | | | |
|---|---|---|---|---|---|---|---|---|
| **Buyer** | **Baled** | **Loose** | **Bundled** | **Grade** | **Rotogravure** | **Colored** | **OCC** | **Grocery Bags** |
| A | X[a] | X | —[b] | #7 | Normal | Normal | — | — |
| B | X | X | — | #7 | — | — | — | — |
| C | NO | X | X | — | No glossy | — | None | X |
| D | NO | X | X | #6 | — | — | — | X |
| E | X | — | — | — | — | — | — | — |
| F | X | — | — | — | — | — | — | — |

*Source:* New York City Department of Sanitation.[20]

[a] X = acceptable.
[b] — = not specified.

**TABLE 8.17 Examples of Buyer Specifications for PET**

| | | | | | Contamination | | | | | |
|---|---|---|---|---|---|---|---|---|---|---|
| **Buyer** | **Baled** | **Granulate** | **Clear** | **Green** | **Caps** | **Labels** | **Ferrous** | **HDPE** | **PVC[a]** | **Other** |
| A | 15 lb/cu ft | No | —[b] | — | X[c] | X | — | — | No | 2% maximum other plastics, metals, paper |
| B | 3 × 4 × 5 ft, 10 lb/cu ft | No | — | — | X | X | — | Mix ok | No | — |
| C | Color sort | X | X | X | — | — | — | — | — | — |
| D | Maximum density | X | X | X | — | — | — | — | — | 3% maximum contamination |
| E | X | — | — | — | — | — | — | — | — | — |
| F | — | X | — | — | — | No | No | — | — | No bottle bottoms |

*Source:* New York City Department of Sanitation.[20]

[a] PVC = polyvinyl chloride.
[b] — = not specified.
[c] X = acceptable.

TABLE 8.18 Examples of Buyer Specifications for HDPE

| | | | | | Contamination | | | | | | |
|---|---|---|---|---|---|---|---|---|---|---|---|
| Buyer | Milk | Nonmilk | Baled | Granulate | Caps | Ferrous | PP[a] | PET | PVC[b] | Moisture | UV[c] Degraded |
| A | X[d] | —[e] | Separate or mix[f] | No | X | — | — | Soda | No | — | — |
| B | X | X | Separate | No | No | No | — | — | — | — | — |
| C | X | X | Separate | X | — | — | No | No | No | Low | Low |
| D | — | — | 700–800 lb/bale | X | No | — | — | — | — | — | — |
| E | — | — | No | X | — | — | — | — | — | — | — |
| F | X | X | X | No | — | — | — | — | — | — | — |

*Source:* New York City Department of Sanitation.[20]

[a] PP = polypropylene.
[b] PVC = polyvinyl chloride.
[c] UV = ultraviolet.
[d] X = acceptable.
[e] — = not specified.
[f] with PET.

2. No. 8 newsprint would be marketable to existing newsprint and construction product industries, and these industries would be willing to increase the percentage of recycled content of their products at the right price.
3. The paperboard industry could consume lower quality (No. 6) newsprint from the facility.

Based on these findings, it was determined that the facility would target marketing 75% of its newsprint as No. 8 news (at $20/ton) and 25% as No. 6 news (zero revenue).

The above example focused on maximizing the production of a high-quality product, while allowing for flexibility.

## REFERENCES


1. Mays, D.A. and P.M. Giodano, ''Landscaping Municipal Waste Compost,'' *BioCycle*, 30(3):37–39, March (1989).
2. Schertz, D.L. ''Conservation Tillage: An Analysis of Acreage Projections in the United States,'' *J. Soil Water Cons.*, 43(3):256–263, May-June (1988).
3. ''Improving Soils with Organic Waste,'' USDA, report prepared in response to Food and Agriculture Act of 1977 (PL 95-113) (1978).
4. Gotaas, H.B. *Composting* (Geneva: World Health Organization, 1956).
5. ''Compost Classification/Quality Standards for the State of Washington,'' Final Report, Prepared by CalRecovery, Inc., for Washington Department of Ecology, September (1990).
6. Eggerth, L.L., L.F. Diaz, and S. Gurkewitz. ''Market Analysis for Multi-Compost Products,'' *BioCycle*, 30(5):29–34, May (1989).
7. Eggerth, L.L. and L.F. Diaz. ''Compost Marketing in the United States,'' presented at BIOWASTE '92, Herning, Denmark (June 1992).
8. ''Market Development Study for Compost,'' prepared by Franklin Associates Ltd., Cal Recovery Systems, Inc., and NUS Corporation for the U.S. Environmental Protection Agency, EPA Contract No. 68-01-7310, June (1990).
9. Robinson, D.W. and J.G. Lamb. *Peat in Horticulture* (New York, Academic Press, 1975).
10. Chaney, R.L. ''The Establishment of Guidelines and Monitoring System for Disposal of Sewage Sludge to Land,'' *Proc. of the Int. Symp. on Land Application of Sewage Sludge*, Tokyo, October (1982).
11. Sharma, R.P. ''Plant-Animal Distribution of Cadmium in the Environment,'' in *Cadmium in the Environment, Part 1, Ecological Cycling*, Nriagu, J.O., Ed. (New York: John Wiley & Sons, 1980).
12. Mennear, J.H., Ed. *Cadmium Toxicity* (New York: Marcel Dekker, 1978).
13. Golueke, C.G. ''Epidemiological Aspects of Sewage Sludge Handling and Management,'' *Proc. of the Int. Symp. on Land Application of Sewage Sludge*, October (1982).
14. ''Portland Area Compost Products Market Study,'' prepared by Cal Recovery Systems, Inc., for Metropolitan Service District, Portland, OR (1988).

15. Cooper, R. et al. *Effect of Disposal Diapers on the Composition of Leachate from a Landfill*, SERL Report 74-3, Sanitation Engineering Research Laboratory, University of California, Berkeley (1974).
16. Diaz, L.F. et al. *Public Health Aspects of Composting Combined Refuse and Sludge and of the Leachates Therefrom*, College of Engineering, University of California, Berkeley (1977).
17. *Anon.*, "Solid Markets for Commercial Compost," *BioCycle*, 23(1):25–26, January-February (1982).
18. "Composting at Johnson City," Final Report on Joint USEPA-TVA Composting Project, Volumes I and II, U.S. EPA, EPA/530/SW31s.2 (1975).
19. Finelli, A. J, "Secondary Materials Markets: A Primer," *Solid Waste & Power*, IV(4):48–56, August (1990).
20. "Recyclables Market Assessment for New York City," prepared by CalRecovery, Inc., for the New York City Department of Sanitation, New York (March 1992).
21. American Paper Institute, *1990 Annual Statistical Summary Waste Paper Utilization*, 5th ed. (New York: API, 1991).
22. Shreve, R.N. and J. A. Brink, Jr. *Chemical Process Industries*, 4th ed. (New York: McGraw Hill, 1977).
23. Savage, G.M. "Recycling of Plastics," presented at the Conference on Achieving Market Expansion Through Plastics Recycling, Sponsored by the Institute for International Research, Miami (September 1989).
24. "Handbook: Material Recovery Facilities for Municipal Solid Waste," prepared by Peer Consultants and CalRecovery, Inc., for the U.S. Environmental Protection Agency, EPA Report No. 625/6-91/031, Cincinnati, OH (September 1991).

# CHAPTER 9

# Biogasification

## BACKGROUND AND STATUS

Biogasification is attracting attention in waste management because in addition to being a treatment method, it also produces a combustible gas, namely, methane. Methane production makes the biogasification process a biological means of converting waste into energy. The fact that biogasification occurs in landfills and produces recoverable methane has intensified the attention accorded the process in the last decade or two.[1]

### Developing Countries

In developing countries, awareness of the energy feature has been manifested by a proliferation of proposed biogasification schemes — particularly in countries deficient in energy sources.[2-4]

Characteristics common to many of the schemes proposed for developing nations are (1) simplicity of design; (2) strong reliance on local skill and use of "native" raw materials for fashioning the necessary containment structures; and (3) very little sophisticated equipment. Only very few of the many proposed designs are realistic and have a chance of functioning as hoped for by the designer.[5,6]

Along with the proliferation of proposed designs has come an escalation in the number of biogasification installations supposedly in operation in various developing countries. Although it is true that the number of constructed installations is not small, there is little evidence to support the magnitude of the reported numbers. Unfortunately, the reports usually fail to mention the significantly large number of unsuccessful ventures into biogasification. A large majority of the failures are due to inappropriate construction and the lack of maintenance.

### Developed Countries

It is a well-known fact that biogasification has been a key element of wastewater treatment in developed nations almost from the beginning of formal wastewater (sewage) treatment.[7-9] It continues to have that status in most modern wastewater treatment systems. The role was and is primarily as a means of treating the settleable (suspended) solids in the sewage stream and the solids produced in secondary and tertiary wastewater treatment (e.g., activated sludge floc). In recent years, it has become a means of treating many industrial wastes. However, in industrial waste treatment, the anaerobic decomposition aspects of the process supersedes the interest in the methane production aspect; although it is by no means ignored.

The attention accorded biogasification of solid waste — manures, agricultural wastes, and the biological decomposable fraction of MSW — has not equalled that in sewage treatment.

The occurrence of the "energy shortage" in the 1970s and the present interest in conservation of energy has occasioned some growth in the interest in the biogasification of solid waste. In the 1970s and early 1980s it aroused a flurry of interest in the agricultural community as a means not only of treating agricultural wastes but also as a means of meeting some of the energy demands of modern farming. However, with the so-called alleviation of the shortage of energy of fossil origin, the interest rapidly waned with respect to waste treatment and to energy production.

With one or two notable exceptions, the interest in biogasifying MSW remained very slight.[10-12] However, in the 1980s a gradual increase in the process as an adjunct or even as alternative to aerobic composting began to develop. The interest was stimulated by the appearance of the "high solids" anaerobic digestion approach.[13,45] "High-solids" anaerobic digestion even has been labeled "anaerobic composting" — in which guise it has been regarded as a serious alternative to aerobic composting in the 1940s and early 1950s.[15-17]

"High-solids" anaerobic digestion differs from "conventional" anaerobic digestion in that the solids content of digesting sludge is on the order of 20% or higher; whereas in conventional digestion, it usually is lower than 3 or 4%, i.e., it is a slurry.

In the sections that follow, a realistic evaluation is made of biogasification as a combination of treatment method and practical source of energy. Important elements of the evaluation are a description and discussion of the basic principles of the process — biological, constructional, and operational design. The presentation of fundamentals is followed by a discussion on the status of the process and on its advantages and limitations.

## PRINCIPLES

### Definition

Depending upon the writer, "biogasification" may alternatively be designated as "methane fermentation," "methane production," or as "anaerobic digestion." The term "methane fermentation" is ambiguous in that literally it also refers to the fermentation of methane, i.e., the *destruction* of the gas through microbial fermentation. Anaerobic digestion is not necessarily attended by methane production. All three terms are used interchangeably in this presentation. In keeping with present terminology and with the purposes of this book, as popularly accepted and for the purposes of this book, "biogasification" is defined as the biological decomposition of organic matter of biological origin under anaerobic conditions with an accompanying production primarily of methane ($CH_4$) and carbon dioxide ($CO_2$). The two features that distinguish the process as defined from other biological decompositions are "under anaerobic conditions" and "the production of methane." In this presentation, the organic material of concern is solid waste.

## Microbiology

A wide variety of microflora is involved in the overall biogasification process. Attempts to identify the key groups have been numerous, but none has met with a completely successful outcome. The reasons for the less-than-successful outcomes are of the same nature as those offered in the chapter on composting and, hence, need not be discussed in this chapter.

Conventionally, biogasification is considered to take place in two distinct yet very closely related and interdependent steps or stages. The two are the "acid phase" ("acid-former phase") and the "methane production phase" (methanogenic phase). According to some researchers, the process is more properly divided into three stages when the substrate is a waste. They precede the acid and the methane stages by one which they term "polymer breakdown." Dividing the process into three stages probably more accurately reflects the microbiology of the overall process. However, a two-stage division makes for simplicity of description and reference and is the one more commonly used in the literature. Moreover, the acid former/methanogen division is justified if one regards the process in terms of premethane activity and methane-forming activity. The two (or three) steps are distinct because they can be separated from each other in terms of reactions and microfloral composition and are developed in sequence. A schematic diagram of the relationship between the three stages is presented in Figure 9.1.

In the biogasification sequence and assuming the three-stage division, the entire process begins with the polymer stage which paves the way for the acid stage, which in turn provides the raw material for the methane stage. Thus, the three stages are dependent one upon the other because the overall process rests upon the maintenance of a relatively critical balance between the respective activities of the three. An imbalance reduces the efficiency of the overall process and may lead to the complete cessation of all microbial activity and, hence, no methane production.

In practice, the traditional division into two phases is more readily apparent than the three-phase division. Thus, when a new culture is started, or as popularly expressed, "when a digester is started," the sequence of readily observable

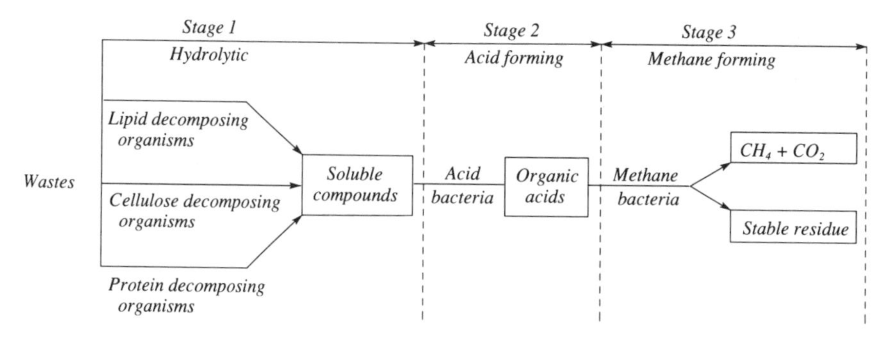


**Figure 9.1.** Relationship of three stages in biogasification.

reactions is a gradual decline in pH level (the acid stage) followed by a similarly gradual rise in pH level and eventually by the production of a gas rich in methane (the methane production stage).

## *Continuous vs Batch Cultures*

An important point to keep in mind is that in an on-going digester operated on a continuous or almost continuous basis, the two (or three) stages are taking place side by side, because each input of waste must go through the two (or three) stages successively. A continuous culture is one to which fresh substrate is added continuously and an equivalent amount is withdrawn. In biogasification practice, feeding and withdrawal once each day or two is considered as being sufficient to justify the designation "continuous." In contrast, a "batch culture" is one to which no material is added and from which none is removed after cultural conditions have been established. In other words, the digester is filled to capacity with a waste to which an "inoculum" may or may not have been added. Thereafter, no material is added or withdrawn from the digester until the digestion process has been completed.

Because in a digester culture operated on a continuous basis, new material is added before material already in the digester has passed through all stages. Some may be as yet in the acid stage, others in transition from the acid to the methane stage, and so on. Inasmuch as all stages are occurring simultaneously, acids are used by the methanogens as rapidly as the acids are synthesized. Consequently no sustained accumulation of acids should occur and the only external manifestation should be the production of methane gas, which may be collected. The methane phase is externally manifested, because the gas leaves the culture and is collected.

The three breakdown stages are discussed individually and in some detail in the succeeding three sections.

*Polymer Stages.* In the polymer breakdown stage, organic wastes are acted upon by a group of facultative microorganisms that enzymatically hydrolyze the polymers of the raw waste into soluble monomers. The monomers become the substrate for the succeeding stage, namely, the acid stage.

The bacterial populations active in the polymer stage are primarily those that have enzymatic systems capable of hydrolyzing the complex molecules of the intact waste particles. Molecules to be hydrolyzed are mainly those of carbohydrates. Others in lesser amounts are those of lipids and proteins. The carbohydrates are represented chiefly by cellulose and other components of plant fiber, such as lignin and hemicellulose. The bacteria involved in hydrolyzing these compounds must as a group possess cellulytic, lipolytic, and proteolytic enzymatic capacity. Because the largest fraction of almost all residues is cellulosic, it becomes apparent that cellulytic activity is the most critical in reducing complex wastes into soluble organic substances.

As it happens in a number of biological waste processing systems, the rate of breakdown of cellulose (or of lipids and proteins) is substantially greater when a mixed colony of hydrolytic bacteria is involved than when a monoculture (pure cultures) is concerned. The reason is in the synergistic action that occurs when many types of microbes are involved. The beneficial effect of the synergism perhaps is the removal of the by-products of the cellulytic bacteria by the noncellulytic forms.

*Acid Stage.* The soluble breakdown products from the first stage are the substrates for the acid-forming organisms in the acid stage. They are converted into short-chain organic acids such as acetic, proprionic, valeric, and lactic acids. Of the acids, acetic acid is the most abundant. Some carbon dioxide also is formed.

As a group, the acid formers are vigorous in growth and are quite tolerant in terms of environmental conditions. The latitude of tolerance perhaps is a collective function of the diversity of component species rather than of an unusual tolerance by any single species. The upshot is that it is only rarely that a slow growth of the acid formers constitutes the limiting factor in an ongoing methane production system. However, in another respect the acid formers can become a very serious limiting factor. This takes place when the synthesis of organic acids by the acid formers is not counterbalanced by the utilization of the acids by methanogens, and to some extent by other organisms. In the absence of such utilizations an acid buildup occurs to the extent that the hydrogen ion concentration (pH) reaches a sharply inhibitory level.

*Methane Stage.* In the methane stage, methanogens convert short-chain fatty acids, $CO_2$, and $H_2$ into methane ($CH_4$) and carbon dioxide ($CO_2$).

In this step, methane is produced in two ways: (1) the organisms involved can break down organic acids (e.g., acetic acid) to methane and carbon dioxide; (2) they can reduce carbon dioxide to methane through the use of hydrogen or formate produced by other bacteria. The end products of the methane stage are methane, carbon dioxide, trace gases, and a stable residue.

### *Cultural Requirements of the Three Stages*

Methanogens grow slowly and are relatively fastidious in their nutritional and environmental requirements. They can only utilize simple organic compounds as nutritional sources, and consequently, in the digestion of wastes they are dependent upon the activities of stages 1 and 2 to provide nutrients in a usable form. In addition to the need for the carbon in the acids produced in stages 1 and 2, they depend upon the nitrogen in the ammonia produced by the breakdown of organic nitrogen compounds to meet this nitrogen requirement.

The sensitivity of the methanogens to key environmental factors is much more acute than that of the acid-forming bacteria. The difference is especially pronounced with respect to pH level. Thus, the tolerated pH range of methanogens is from pH 6.5 to 7.5, and the optimum is 7.0, whereas the tolerated range for the acid-forming group is from pH 4.5 or 5.0 to 7.5 or even 8.0. Another important difference is the sensitivity of the methanogens to atmospheric oxygen ($O_2$). Methanogens are obligate anaerobes, and traces of $O_2$ inhibit their growth. Even highly oxidized compounds, such as nitrites or nitrates, can inhibit the growth of the microorganisms. On the other hand, most of the acid formers are facultative anaerobes, i.e., the presence of $O_2$ is not inhibitory to their growth.

The practical import of the preceding paragraphs on microbiology can be summarized in terms of the potential of the major microbial groups to lead to the development of rate-limiting factors. The first potentially rate-limiting factor is the conversion of insoluble celluloses by extracellulases into soluble carbohydrates, as well as the solubilization of complex organic nitrogenous compounds. The second is the activity of acid forming bacteria in converting the soluble carbohydrates into low weight fatty acids. The third factor is the conversion of acids by the methanogens into $CH_4$ and $CO_2$. The third factor, conversion into methane, is generally regarded as the rate-limiting stage for the process as a whole because it is the final step and because the methanogens are basically slow growing.

## Environmental Factors

The types of environmental factors that determine the course of the digestion process are those that affect all biological processes. With respect to anaerobic digestion, the more important factors are temperature, pH level, oxidation-reduction potential, and substrate. The maintenance of a low oxidation-reduction potential is essential to the continued activity of the process. The disastrous effect of any elevation of the level due to the presence of $O_2$ or of a highly oxidized material has already been mentioned. The importance of maintaining a suitable pH level (i.e., 6.5 to 7.5) also has been stressed in a preceding section.

### *Temperature*

Temperature directly influences the rate of digestion, as is manifested by amount of gas produced and volatile solids converted. Increase in rates of gas production and destruction of volatile matter directly reflect rise in temperature until a level of 38 to about 42°C is reached. If the rise in temperature continues beyond 42°C, the rates begin to decline gradually until the temperature reaches 45°C. Thereafter, the rates drop temporarily abruptly and drastically until the culture adapts to the higher temperature range. Attainment of adaptation is marked by a resumption of the upward trend at about 50°C. The upward trend continues until a second plateau is reached at 55°C. Rates drop drastically after a temperature of 65°C has been passed.

An important point not to be overlooked is that a culture adapted to growth at mesophilic temperatures (5 to 42°C) will rapidly perish upon being suddenly exposed to thermophilic temperatures (45°C and higher). However, the culture can be adapted, i.e., converted into a thermotolerant or even thermophilic culture either through enrichment or through adaptation. Consequently, to operate a digester under thermophilic conditions, either an existing culture of thermophiles must be used, or one must be developed. Development is a time-consuming operation and depends upon the chance occurrence of "wild" strains of thermophiles in the "starting" culture.

### *Enrichment Culture and Adaptation*

Two approaches for developing a thermophilic culture are as follows. In one, the culture is set up and then is exposed to a gradually rising temperature — beginning at 35°C and eventually (30 days or longer) reaching 50 to 55°C. The second approach is more drastic. The digester culture is immediately exposed to thermophilic conditions. Although the culture thereupon seemingly will have been destroyed, if left undisturbed long enough, it eventually will resume activity. In the drastic approach, all obligate mesophiles are killed off and only a few isolated thermophiles, obligate or facultative, remain. These latter gradually multiply until the requisite population density is reached.

The time required for developing a thermophilic microbial population should be an important consideration in decisions regarding the advisability of using a thermophilic system. If a mesophilic culture fails, a new one is more readily established than is a thermophilic culture. A probably more serious reason against the use of a thermophilic system is the added energy input needed to maintain the substantially higher temperature. Any gain in gas production and destruction of volatile matter and the shortening of the detention period possible with thermophilic digestion must be weighed against the increase in energy requirements. That is, reductions in capital costs may not be compensated by costs due to operating and maintenance.

## Substrate

The waste to be digested forms the principal substrate for the digester culture. The qualification "principal" is used because certain wastes may be deficient in one or more elements essential to the nutrition of the microbial population. Under such a circumstance either another waste must be added that contains the missing nutrients or an appropriate inorganic source must be supplied. Operations in which inorganic nutrients are added are rare, if for no reason other than due to the cost involved.

The substrate may be considered from these three aspects: physical characteristics, chemical composition, and biodegradability. Strictly speaking, biodegradability is partly a function of chemical composition, i.e., of the molecular structure of the components of the waste. Hereinafter, the substrate is referred to as "digester feed" or "input."

### *Physical Characteristics*

The two physical characteristics of the feed that are of principal concern are particle size and moisture content.

*Particle Size.* As stated in the chapter on composting, rate of bacterial attack is in part a function of the ratio of surface area to mass. The higher the ratio, the faster the rate of decomposition. This ratio is especially important in the digestion of municipal refuse and the more fibrous of the agricultural wastes. On the other hand, the particle size of highly putrescible wastes or of food preparation wastes need not be as small. Manures of cattle and fowl can be added directly to a digester without adverse effect unless they are intermingled with a substantial amount of bedding.

*Moisture.* For conventional digestion, the moisture content should be such that the feed is in slurry form. Experience indicates that a solids content of 5 to 8% (i.e., moisture, 92 to 95%) is appropriate. Too high a solids content leads to inadequate mixing, with the objectionable consequences to be described later. Too low a solids content involves a digester volume larger than would normally be necessary. Because of the expense involved, digester size can be the deciding factor regarding economic feasibility.

"High Solids" Digestion. The high-solids digestion that appeared in the past one or two decades involves solids concentrations that range from about 10 to about 15% (moisture, 85 to 90%). In terms of biological limits, the upper solids range for anaerobic decomposition roughly parallels that for composting. However, practicality and the current use of the term "digestion" dictate the 10 to 15% range for the present.

### *Chemical Composition*

Two important factors regarding chemical composition are the elemental composition of the feed and the structure of the molecules that contain the elements.

The elements required for the successful functioning of all biological systems in general also are essential in anaerobic digestion. The macronutrients are nitrogen, carbon, and phosphorus. The micronutrients are the usual trace elements such as sodium, cobalt, manganese, and several other metallic elements.

When wastes constitute the feedstock, trace elements generally are present in sufficient concentration.

*C to N Ratio.* Problems can arise when there is an imbalance between the carbon and the nitrogen contents of the feed. In other words, an appropriate ratio of carbon to nitrogen (C:N) is a requisite for the continued successful functioning of a digester. When the C:N is too high, acids accumulate and methane production drops. Apparently, the acid formers are more efficient in assimilating nitrogen than are the methane formers.

When the C:N is too low, more nitrogen is converted into ammonium-nitrogen than can be assimilated by the methanogens. Consequently, ammonia reaches concentrations that are toxic to the microbes. The level at which the ratio is suitable depends in part upon the nature of the substrate. If the substrate is readily or moderately degradable, i.e., the carbon and nitrogen are readily available to the bacteria, the uppermost C:N is on the order of 25 to 30:1. When the feed is largely refractory to bacterial breakdown, i.e., the carbon and nitrogen are difficultly available to the bacteria, the C:N can be as high as 35:1 or even 40:1. Examples of refractory wastes are wood and other lignaceous wastes, such as rice hulls and straw.

Unless pretreated by exposure to heat, pressure, and acid or alkali, wood breaks down so slowly that anaerobic digestion becomes unfeasible in practice. Because of their tendency to float, wood, straw, rice hulls, and other wastes of low density are unsuitable for digestion. In practice, low-density wastes aggravate the formation of the scum (foam) that leads to the development of numerous problems in anaerobic digestion. Paper other than newsprint breaks down readily. Newsprint is an exception because its constituent cellulosic fibers are partially masked by lignin, a substance that is very resistant to microbial attack.

*Nutritional Characteristics of Representative Wastes.* Chemical and other characteristics of some representative wastes are described at this point so as to assist in decisions regarding the feasibility of a prospective biogasification project. The information in Table 9.1 can be used in judging the biodegradability of the more common crop and forest residues. The relative digestibility of the tabulated wastes can be ascertained by comparing the ratios of water-soluble constituents to the combined lignin-cellulose contents. The higher the ratio, the more readily is the material digested. With the exception of newsprint, paper is digested relatively rapidly when present with an appropriate nitrogen source. Newsprint, being almost entirely cellulosic and lignaceous, would normally be slowly digestible. The nitrogen content and carbon-nitrogen ratios of several wastes are listed in Table 9.2.

*Toxic Substances.* The term "toxic substances" can be interpreted in many ways — ranging from substances that are lethal or sharply inhibitory for all microbes to those that become lethal or inhibitory only after a critical concentration is reached. The concentration varies according to the substance and the

**TABLE 9.1 Chemical Composition of Common Plant and Forest Residues[29] (% air-dry material)**

| Constituent | Young Rye Plant | Mature Wheat Straw | Soybean Tops | Alfalfa Tops | Young Cornstalks | More Mature Cornstalks | Young Pine Needles | Old Pine Needles | Oak Leaves, Green | Oak Leaves Mature, Brown |
|---|---|---|---|---|---|---|---|---|---|---|
| Fats and waxes | 2.35 | 1.10 | 3.80 | 10.41 | 3.42 | 5.94 | 7.65 | 23.92 | 7.75 | 4.01 |
| Water-soluble constituents | 29.54 | 5.57 | 22.09 | 17.24 | 28.27 | 14.14 | 13.02 | 7.29 | 22 .02 | 15.32 |
| Hemicelluloses | 12.67 | 26.35 | 11.08 | 13.14 | 20.38 | 21.91 | 14.68 | 18.98 | 12.50 | 15.60 |
| Cellulose | 17.84 | 39.10 | 28.53 | 23.65 | 23.05 | 28.67 | 18.26 | 16.43 | 15.92 | 17.18 |
| Lignin | 10.61 | 21.60 | 13.84 | 8.95 | 9.68 | 9.46 | 27.63[a] | 22.68[a] | 20.67[a] | 29.66[a] |
| Protein | 12.26 | 2.10 | 11.04 | 12.81 | 2.61 | 2.44 | 8.53 | 2.19 | 9.18 | 3.47 |
| Ash | 12.55 | 3.53 | 9.14 | 10.30 | 7.40 | 7.54 | 3.08 | 2.51 | 6.40 | 4.68 |

[a] The high lignin content is partially an artifact due to the analytical procedure used in determining it.

**TABLE 9.2 Approximate Nitrogen Content and C:N of Several Waste Materials[30,31]**

| Material | Total-N (% dry wt) | C:N |
|---|---|---|
| Animal wastes | | |
| Urine | 15–18 | 0.8 |
| Blood | 10–14 | 3 |
| Fish scraps | 6.5–10 | 5.1 |
| Mixed slaughterhouse wastes | 7–10 | 2 |
| Poultry manure | 6.3 | — |
| Sheep manure | 3.8 | — |
| Pig manure | 3.8 | — |
| Horse manure | 2.3 | 25 |
| Cow manure | 1.7 | 18 |
| Farmyard manure (average) | 2.15 | 14 |
| Night soil | 5.5–6.5 | 6–10 |
| Plant wastes | | |
| Young grass clippings (hay) | 4.0 | 12 |
| Grass clippings (average mixed) | 2.4 | 19 |
| Purslane | 4.5 | 8 |
| Amaranthus | 3.6 | 11 |
| Cocksfoot | 2.6 | 19 |
| Lucerne | 2.4–3.0 | 16–20 |
| Seaweed | 1.9 | 19 |
| Cut straw | 1.1 | 48 |
| Flax waste (phormium) | 1.0 | 58 |
| Wheat straw | 0.3 | 128 |
| Rotten sawdust | 0.25 | 208 |
| Raw sawdust | 0.1 | 511 |
| Household wastes | | |
| Raw garbage | 2.2 | 25 |
| Bread | 2.1 | — |
| Potato tops | 1.5 | 25 |
| Paper | nil | — |
| Refuse | 0.8–2.0 | 25–60 |

group of affected microbes. Substances that become critical only above a certain concentration include those that are essential to the well being of the microbial population at trace concentrations.

The presence of materials toxic to the methanogens is the first to be noticed, simply because gas production drops. On the other hand, although the toxin may be equally inhibitory for the acid-forming bacteria and for the hydrolyzers, the effect is not as apparent as it is upon the methanogens. Among the more common toxins for methanogens are molecular ammonia (i.e., ammonia gas), ammonium ion, soluble sulfides, and soluble salts of metals such as copper, cadmium, zinc, and nickel. Several organic substances also may adversely affect the organisms. Molecular ammonia becomes toxic when the concentration exceeds 1500 to 3000 mg/L of total ammonia-nitrogen at a pH higher than 7.4. The ammonium ion is toxic at concentrations greater than 3000 mg/L of total ammonium-nitrogen at all pH levels. The ammonium can exist in equilibrium with dissolved ammonia gas as follows:

$$NH_4^+ \rightleftarrows NH_3 + H^+$$

The pH level determines the degree of the toxicity of ammonia-ammonium because of its effect on the equilibrium between the two forms. The equilibrium shifts towards the ammonium ion at low pH levels and inhibition begins at 3000 mg/L. At the higher pH level, the shift is toward ammonia gas, and inhibition may begin at 1500 mg/L. Despite the possibility of ammonia becoming toxic, it should not be forgotten that ammonium-nitrogen is one of the substances that is essential to microbial nutrition in biogasification.

Soluble sulfides become toxic when their concentrations approach 50 to 100 mg/L. The soluble metal salts become toxic when their concentrations exceed a few parts per million.

Sodium, potassium, calcium, magnesium salts, and other alkali and alkaline-earth metals salts are stimulatory at one concentration, and inhibitory at another. The toxicity is a function of the cation fraction of the salt. Sodium is stimulatory at concentrations of 100 to 200 mg/L; potassium, at 200 to 400 mg/L; calcium, at 100 to 200 mg/L; and magnesium, at 75 to 150 mg/L. Sodium begins to be inhibitory at concentrations approaching 3500 mg/L; potassium and calcium, at 2500 mg/L; and magnesium, at 1000 mg/L.

Because materials exert their toxic effect only when in solution, any procedure that renders them insoluble will have an ameliorative effect. For example, the toxicity of a soluble sulfide present in an excessively high concentration can be lessened by changing the pH of the culture (i.e., of the suspending medium) or by adding a heavy metal to act as a precipitant. The latter recourse should be followed sparingly because the metals lower the agricultural utility of the sludge. Conversely, a poisoning of the microbial population due to an excessive concentration of heavy metals can be countered by adding a soluble sulfide. Complexed heavy metals may be present in concentrations of as much as 100 mg/L without interfering with the activity of the culture. A precaution to be taken in such a case is to add the sulfide only to the extent needed to precipitate the heavy metals. Another factor — unrelated to the well being of the culture — is the fact that the precipitated metals become an integral part of the sludge and consequently must be considered when disposing of the sludge.

*Transfer of Dissolved Products.* A limiting factor suggested by Finney and Evans[32] is the rate of transfer of dissolved metabolic and other products from the liquid to the gaseous phase. They reason that as the biological retention time is shortened, i.e., the developmental period abbreviated, individual bacterial cells must increase their rate of reproduction. A consequence of the accelerated reproduction is an increase in gas production and the need to transport metabolic products away from the vicinity of individual cells. Inhibition arising from an inadequate transfer from liquid to gaseous phase may occur when the bacteria individually are completely surrounded by a wall of bubbles, as would occur at a very high substrate concentration. The envelope of bubbles interferes with diffusion of substrate into intracellular spaces. A solution to the problem is to vigorously agitate the culture, as for example, by thoroughly mixing on a continuous basis.

## OPERATION

The operation of a gasification facility can be described and discussed under three aspects: (1) parameters, (2) procedures, and (3) corrective measures. Although strictly speaking, corrective measures are procedures, they are treated separately in this book for the sake of convenience and clarity.

### Parameters

The principal parameters on which digester performance is judged are gas production and composition, rate and extent of volatile solids destruction, alkalinity, volatile acid content, and pH level. Although a range of values for the preceding parameters has been developed for the digestion of sewage sludge, the values are not always directly applicable to the digestion of solid wastes, principally because of differences between the wastes in terms of chemical and physical structure. In some instances, parameter values considered to represent optimum performance reflect those of optimum environmental conditions. Two such parameters are pH and alkalinity. Other parameters are based upon efficiency of digester performance in terms of destruction and conversion of the input wastes. Gas production and destruction of volatile solids are two examples. (For reasons given in an earlier chapter, the term ''volatile solids'' refers to organic solids.)

#### *Gas Production and Composition*

Gas production serves both as an indicator of the condition of a culture and of the efficiency of digester performance. Gas production usually is expressed in terms of volume of gas produced per unit mass of total solids and/or of volatile solids introduced. Expressing gas production in terms of total solids introduced is useful for predicting total gas production with a given waste. The prediction is direct. Gas production per unit of volatile solids introduced is a good operational parameter in that all input wastes are reduced to a common denominator (namely, volatile solids content) and volatile solids are the only solids converted into gas.

Production per unit of volatile solids may be taken in terms of those introduced and in terms of those destroyed. Thus, the expression is either volume of gas per unit mass of volatile solids destroyed or per unit mass of volatile solids introduced into the digester. The latter is probably the better parameter because it is a measure of how well (efficiently) the culture utilizes the volatile solids introduced into it. Gas production per unit of total solids depends both on the volatile solids content of the total solids and on the extent to which the volatile solids are converted into gas.

The expected gas volume per unit of volatile matter is a function of the detention period and other operational features as well as of the nature of the waste. With raw sewage sludge as the feed, the yield is 0.374 to 0.454 $m^3/kg$ of introduced solids. Yields to be expected from the digestion of other wastes are indicated in Tables 9.3 and 9.4. Because the yields listed in Table 9.3 are based on total solids fed, it would be expected that yields per kilogram of volatile solids would be greater.

As of 1991, reliable data on yields of gas from municipal-scale biogasification of MSW in the U.S. were negligible, because the only experience on such a scale was a demonstration project. Results obtained in the project were inconclusive. Research studies have been many and varied and the reported values are correspondingly diverse.[39-44] The characteristically high C:N (55 or greater) of U.S. MSW makes it essential that the waste be enriched with a nitrogen source so that digestion can progress without the onset of difficulties. Because sewage sludge could be such a source, codigestion of refuse and sewage sludge is receiving increasingly serious consideration.

The production rate to use as a parameter for the digestion of a given waste is the one that is characteristic of the culture after it has reached "steady state." When a new culture is begun, gas production gradually increases from almost zero yield to a certain point, after which it levels off — i.e., steady state is reached. With sewage sludge, the yield upon reaching the steady state would be on the order of the 6.0 to 7.3 cu $ft^3/lb$, as mentioned before. A sharp drop or slow but steady decline from the steady-state yield indicates that the culture is not functioning properly.

Gas production is a particularly useful parameter because it is easily recognized and is an almost immediate response to an unfavorable condition. Here again, attention is called to the daily fluctuation characteristic of most parameters even after a steady state is reached with almost every biological system. However, the fluctuations are only slight and a given deviation is not persistent. The minor deviations are the result of the practical impossibility of maintaining all operational and environmental conditions at a constant level. However, a continued decline of 4 or 5 days of duration could be taken as an indication of trouble. On the other hand, an abrupt, steep drop in yield is a positive sign of an imminent danger to the culture.

Gas yield, when used as a parameter, should not be interpreted as being completely independent of gas composition. The gas of interest in the composition is methane. The methane content to be expected is a function of the composition of the waste fed the digester and of the condition of the methane population. Generally, the greater the carbohydrate concentration in the waste, the more closely will the ratio of methane to carbon dioxide approach 1:1. On the other hand, the fraction as methane may be as much as 65% when raw sewage

**TABLE 9.3 Biogas Production from the Digestion of Various Wastes**

| Raw Material | Biogas/Unit wt. of Dry Solids ($m^3$/kg) | Temp. (°C) | $CH_4$ Content of Gas (%) | Detention Time (days) |
|---|---|---|---|---|
| Cattle manure | 0.20–0.33 | 11.1–31.1 | — | — |
| Poultry manure | 0.31–0.56 | 32.6–50.6 | 58–60 | 9–30 |
| Swine manure | 0.49–0.76 | 32.6–32.9 | 58–61 | 10–15 |
| Sheep manure | 0.37–0.61 | — | 64 | 20 |
| Forage leaves | 0.5 | — | — | 29 |
| Sugarbeet leaves | 0.5 | — | — | 11–20 |
| Algae | 0.32 | 45–50 | 55 | 11–20 |
| Night soil | 0.38 | 20–26 | — | 21 |
| Municipal refuse (U.S.)[a] | 0.31–0.35 | 35–40 | 55–60 | 15–30 |

*Source:* NAS.[6]

[a] Estimated yield per kilogram of volatile solids introduced.

**TABLE 9.4 Animal Manure Production in Terms of NPK Equivalent and of Potential Methane Production[a]**

| Animal | Waste Production (kg/day/animal) | Methane (Msm/day/animal) | N[b] | P[b] | K[b] |
|---|---|---|---|---|---|
| Bovine | | | | | |
| Beef | 38 | 1.19 | 32 | 15 | 40 |
| Dairy | 52 | 1.58 | 64 | 29 | 79 |
| Replacement | 34 | 1.02 | 21 | 10 | 27 |
| Swine | | | | | |
| Sows (136 kg) | 14 | 0.51 | 15 | 8 | 5.4 |
| Hogs (68 kg) | 7.3 | 0.25 | 7.7 | 4.1 | 2.7 |
| Weaners (27 kg) | 3.6 | 0.14 | 4.1 | 1.8 | 1.4 |
| Poultry (per 1000) | | | | | |
| Broilers | 28 | 0.74 | 64 | 27 | 23 |
| Layers | 118 | 3.11 | 499 | 403 | 222 |
| Turkeys | 134 | 3.42 | 245 | 195 | 109 |
| Horses | 17 | 0.48 | 45 | 17 | 35 |

[a] Based on information in Reference 33.
[b] Kilogram per year per animal.

sludge is the substrate. In the early stages of the development of a digester culture, gas that may be produced is mostly carbon dioxide. However, the methane content increases gradually until a relatively constant level is reached. Therefore, a consistent decline in methane content indicates a difficulty. The upshot is that a persistent decline in gas production coupled with a drop in the concentration of methane constitutes very positive evidence of the existence of inhibitory conditions in the digester culture.

## *Destruction of Volatile Matter*

Because digestion is a decomposition process, some volatile matter will be destroyed. Inasmuch as rate and amount of destruction are functions of the degradability of the substrate, this fact should be kept in mind when interpreting a departure from a steady-state value. For example, if the rate of destruction has been on the order of 50 to 60% with swine manure as the substrate, it may well be lower if one day cattle manure were substituted for the swine manure.

The range of destruction to be expected can be as much as 60 to 70% with wastes from food preparation as the substrate to as low as 30 to 40% when newsprint is the principal carbon source. In addition to degradability, contributing factors are temperature and detention period. All three must be taken into consideration when evaluating changes in rate of destruction of volatile matter.

### Volatile Acid Content

Volatile acids become limiting only when they bring about a drop in pH to a level that is inhibitory to methanogens. The concentration required to bring about such a drop ultimately depends upon the buffering capacity of the culture. For example, after a suitable adaptation, methanogens can flourish at volatile acid concentrations as high as 10,000 mg/L.[18] On the other hand, with a poorly buffered culture, the inhibitory concentration may be as low as 200 or 300 mg/L. Therefore, the significance of volatile acid concentration as a parameter rests upon the concentration remaining constant rather than upon a specific concentration.

Because of the lesser vulnerability of acid formers to adverse conditions, the presence of a condition inhibitory to the culture *as a whole* generally is manifested by a gain in concentration of volatile acids. Although some textbooks on sludge digestion state that a concentration of volatile acids at 2000 mg/L is lethal to methane formers, the statement is not always applicable. The practical danger of naming any one concentration or narrow range of concentrations as being the optimum concentration is in the false sense of security that comes from the failure to perceive any danger in an increase in acid concentration, so long as it does not exceed 2000 mg/L. The fact is that after a steady-state concentration has been reached — which could be as low as 200 mg/L — any persistent increase is to be regarded as an indication of danger.

Volatile acids concentration usually is expressed as milligram per liter of acetic acid. Methods of volatile acid determination are given in textbooks on wastewater analysis. A disadvantage of volatile acid concentration as a parameter is in the effort and equipment required in making the analysis.

### Hydrogen Ion Concentration

Hydrogen ion concentration can be regarded as an operational parameter because (1) methanogens are inhibited at pH levels beyond the relatively narrow range of 6.5 to 7.5; and (2) pH level is a manifestation of volatile acid formation. The utility of pH level as a parameter is diminished by the fact that pH level not only is a function of volatile acid concentration, it also depends upon the buffering capacity of the culture. Because of this fact, a culture could be well on the way to complete inhibition before a substantial change could be noted in the pH level. The utility of any parameter depends to a large extent upon the immediacy with which it responds to significant changes in the overall conditions of the culture.

### Buffering Capacity

In sanitary engineering practice, buffering capacity usually is referred to as "alkalinity". The utility of alkalinity as conventionally determined is open to question. A major objection is that the routine alkalinity analysis does not afford the complete information needed for the safe operation of a digester. The routine determination of alkalinity to pH 4.0 spans about 80% of the acetate alkalinity as well as pertinent bicarbonate alkalinity. The difficulty is that the buffering range of acetate is effective only from pH 3.75 to 5.75, a range well below that tolerated by the methanogens. The bicarbonate alkalinity can be obtained by subtracting acetic alkalinity from total alkalinity. The bicarbonate alkalinity required to maintain a pH level at 7.0 depends upon the carbon dioxide content of the digester gas. For example, with the $CO_2$ of the gas at 25%, the required bicarbonate alkalinity would be on the order of 2000 mg/L. On the other hand, the needed alkalinity would be 4000 mg/L, if the $CO_2$ concentration were from 50 to 53%. Typically, in well-operated digesters, the alkalinity total ranges from 1500 to 5000 mg/L as acetic acid.

### Correcting a Malfunctioning Digester

If the values of the various parameters are such as to indicate the onset of unfavorable conditions, and thereby a subsequent deterioration in digester performance, then certain remedial measures must be taken. Of course, before the measures can be taken, cause or causes of the change in performance must be identified. Otherwise, only the symptoms are treated and the underlying causes are neglected. For example, the addition of lime or buffer to elevate a low pH only raises the pH level. It leaves intact the problem responsible for the pH decline. Moreover, the problem does not end with the introduction of lime into the digester. Unless due precautions are taken, the lime may become a cement-like precipitate upon the bottom of the digester.

An alternative method of raising the pH level is to add ammonia. Although ammonia is very effective in neutralizing a digester culture, its use may endanger the culture because relatively small concentrations of ammonia are very toxic to microbes. Yet another approach is to enhance the bicarbonate alkalinity through the addition of sodium bicarbonate. Cost and availability of the bicarbonate may constitute a substantial barrier to its use in small operations.

Until a suitable remedial measure is decided upon, the course with most in its favor, is to cease feeding the digester. Continued feeding only magnifies the problem of remedying the difficulty. The problem of restoring a completely inactivated ("stuck") digester to normal functioning is one for which no clear-cut answer is presently available. If doing so is feasible, the best recourse is to dispose of the digester culture and then to develop a new one. Such an action is fairly easily accomplished if the scale of the digester is small. However, the situation becomes so complicated as to be almost hopeless when a large (municipal-scale) digester is being operated. Regulatory agencies are adamantly opposed to the unrestricted discharge of a digester's contents because of the unfavorable impact upon the environment and land, air, and water resources of such an action.

## Procedures

The procedures described in this section mostly pertain to biogasification in which the feedstock is slurried.

### *Mixing*

Mixing or agitation is of key importance to the efficient operation of a digester in which the contents are slurried. It makes the difference between high-rate digestion and conventional or slow digestion. With certain substrates, such as fibrous materials, for example, it is essential to the continued successful operation of a digester. For instance, many of the unsuccessful digester operations in developing nations can be attributed in part to inadequate mixing. The beneficial influence of mixing on the digestion process is the removal of metabolic waste products and spent medium from the immediate environment of the individual cells and replacing the spent medium with replenished medium.

A second important function of mixing is to counteract sedimentation. If a digester culture is allowed to remain quiescent, it separates into the four layers diagrammed in Figure 9.2.

As is shown in Figure 9.2, the uppermost layer is the scum layer. The scum layer is a mixture mostly of low-density materials (e.g., wood, feathers, straw, chaff) and some fine inerts buoyed by a foamy mass of supernatant. The layer may vary from as little as 2 in. in thickness to as much as 12 in. or more. The foam consists of bubbles of supernatant. The bubbles dissipate very slowly because of the high surface tension of the supernatant. The scum layer lowers the efficiency of the operation because it traps, collects, and isolates degradable material from the active zone of digestion. In the digestion of the fibrous and low-density materials mentioned earlier, the extent of the accumulation of potentially digestible materials in the scum layer can be substantial. The scum layer lowers the efficiency of energy conversion by interfering with the tapping of the energy content of the material accumulated in it.

In addition to the unfavorable impact on cultural efficiency, a sizeable scum layer interferes with operational procedures such as gas collection, recirculation, and the lesser mixing systems. Moreover, a portion of the digester volume serves no purpose, and thus the effective capacity of the digester is lowered.

Scum formation is minimized, if not avoided completely, through mixing. The greater the tendency to scum formation, the more vigorous must be the mixing. If the supernatant has a moderate viscosity and the digesting wastes are not greatly fibrous or rich in low-density materials, adequate mixing can be

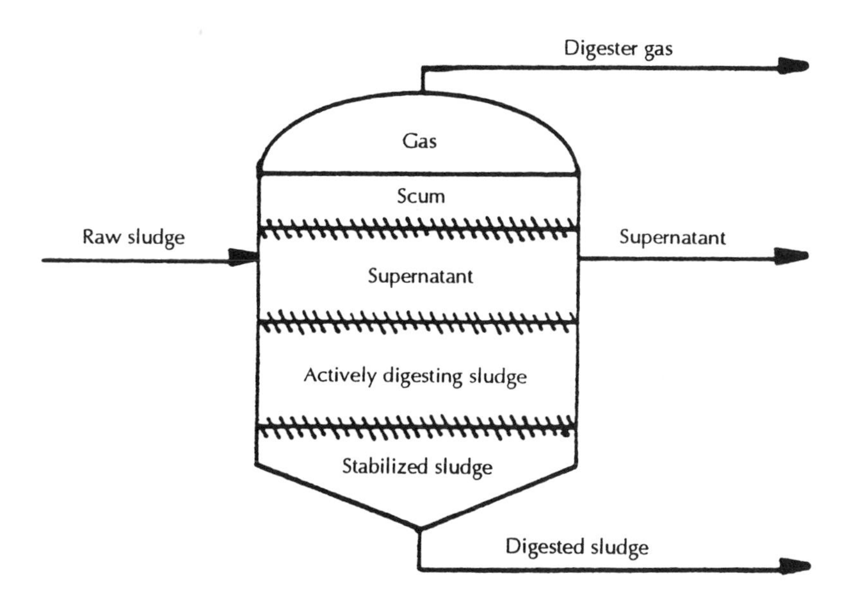


**Figure 9.2.** Layers formed in digesters.

supplied by recirculating either the digester liquid or the digester gas. If the liquid is recirculated, it is taken from the bottom of the digester and introduced above the culture through one or more jets. The converse is done with digester gas. The gas is removed from the gas plenum and is introduced at the bottom of the culture. The bubbles accomplish mixing as they ascend through the culture.

If the tendency to form scum is strong and the scum is persistent, recirculation either of gas or of liquid is not sufficiently vigorous to prevent scum formation. In such a case, the only recourse is mechanical mixing. Mechanical mixing may be accomplished in a variety of ways, but all basically involve the rotation of a paddle or paddles in the culture suspension. The arrangement and location of the paddles may vary. Regarding digesters having a volume less than a cubic meter or two, mixing can be done manually, i.e., the paddle device can be rotated by hand. With larger digesters, the rotation is accomplished mechanically. With digesters up to 106 or 140 $ft^3$ in capacity, the mixing need not be on a continuous basis. Ideally, it should be done once or twice each day, although an occasional lapse of a day or two would not be amiss.

Although theoretically, vigorous agitation enhances the activity of the culture, it also is particularly effective in promoting scum formation. Mixing by recirculation is especially effective in that regard. Consequently, a balance must be made between the ability of the mixing system to break up the scum layer and its tendency to promote the formation of the layer.

### *Loading*

Rate and amount of loading are determining factors in the amount of the energy that is recovered from the wastes and in the efficiency of utilization of digester capacity. The consequence of overloading is a decline in the amount of energy recovered, an eventual complete cessation of microbial activity, and the destruction of the microbial population. In the vernacular, a digester in such a condition is said to be "stuck." On the other hand, if the loading is less than that permitted by the capacity of the digester, digester capacity is used inefficiently. The economic import is the added cost associated with an unnecessarily large unit. It might be said that underloading has the advantage of increasing the percentage of energy recovered. However, the principle of diminishing returns may intervene in that the additional energy recovered may not warrant the extra effort.

As is true of practically every parameter and operational procedure, loading is a function of the nature of the substrate and the degree to which operating conditions approximate the optimum. Loading may be expressed in terms of units of volatile solids introduced per unit of digester capacity per unit of time (usually, "day"), or in terms of total solids per unit of digester capacity per unit of time.

Expressing loading in terms of volatile solids has the advantage of uniformity and a certain degree of universality, because the percentage of volatile solids varies with type of waste. In this discussion, loading is expressed in terms of volatile solids. All loading rates are on the basis of *dry* weight of the solids.

Appropriate loading rates vary with the nature of the waste. Generally, if the waste is readily biodegradable, the loading in terms of volatile solids is less because the fraction of material directly available to the organisms, particularly the acid formers, is greater. On the other hand, the nutrients in refractory organic materials are less readily available to the microorganisms. Therefore, the loadings can be somewhat larger without leading to adverse results. For example, at a 20- to 30-day detention period, the loading rate with raw sewage sludge is on the order of 0.08 to 0.16 $lb/ft^3$ digester volume per day. With a 1:1 mixture of sewage sludge and organic refuse rich in paper as is the type generated in the U.S., the loading can be as much as 0.2 $lb/ft^3$. Reported loading rates with cow manure as the feed range from 0.07 to 0.33 $lb/ft^3/day$.[6] A good compromise seems to be 0.23 $lb/ft^3/day$ (detention time, 17 days).

### Detention Time

Synonyms for "detention time or detention period" are "retention time" and "residence time."

As generally used, the term "detention time" is used only in reference to continuous cultures. With a typical continuous culture, the microbial population (suspended solids) may be placed on one detention period and the liquid phase on another; or the two may be placed on a common detention time. The liquid and the common detention periods are referred to as the hydraulic detention period. Hydraulic detention time can be expressed as

$$t = V/Q$$

where t is the detention time, V, the culture volume, and Q, the throughput per unit of time. In large-scale digestion, the hydraulic detention period is the one usually applied, and it undoubtedly is better suited to small-scale anaerobic digestion.

The dual detention period is particularly appropriate for any one of the following conditions: (1) the microbial mass constitutes the bulk of the settleable solids; (2) the microbial growth rate is so rapid that water consumption would be excessive; and (3) the hydraulic detention time is the only one applied. On the other hand, the rapidity of the rate of nutrient depletion with a rapidly growing culture may be such that an abbreviated detention period would be warranted. The activated sludge method of wastewater treatment is an example of an aerobic system that involves separate microbial and hydraulic detention times. An argument against a dual detention period is the additional handling involved and the chances of exposing the methanogens to atmospheric oxygen. Another reason could be poor settling characteristics and a substantial concentration of inert fines. With such a combination, the net effect would be a gradual accumulation of inert fines without an accompanying recirculation of microbes. Therefore, the remainder of this section deals only with the common (or single) hydraulic detention period.

Applying the proper detention time is important for several reasons, among which are the following: (1) if the detention time is too long, the digester capacity is not used efficiently and the digester is too large; (2) an excessively lengthy detention period may allow the bacterial population to proceed beyond the phase of exponential multiplication (in other words, the average age of the microbial population is such that the organisms have passed their period of peak capacity); and (3) if the hydraulic detention period is too short, the rate of bacterial multiplication may not be sufficient to compensate for the bacteria discharged in the effluent. Eventually the culture is "washed out." Even were the culture able to maintain itself, the population would be less than adequate for accomplishing the required energy conversion. In effect, the full amount of potentially available energy would not be recovered.

From the preceding two paragraphs, it becomes evident that the ideal detention time is one in which (1) the microbial population, particularly that of the methanogens, is maintained in the exponential growth phase; and (2) the greater part of the reclaimable energy in the waste is converted into the chemical energy of methane.

The proper detention period is a function of a collection of environmental and operational factors and of the composition of the substrate. The more closely conditions approach optimum and the more decomposable the waste, the shorter can be the detention period. The ultimate limitation, all conditions being optimum, is the *genetic* makeup of the bacteria. In anaerobic digestion, it is the genetic makeup of the methanogens that in practice makes it necessary to apply detention periods in terms of weeks rather than of hours.

Unfortunately, the trend is to unduly shorten detention times. For typical U.S. refuse, a 15-day detention period probably would be satisfactory and a 30-day period might verge on being too long. However, it should be pointed out that researchers dealing with pure cellulose (reagent grade) suspended in a medium suitably fortified with nutrients found that the cellulose could be digested at a 5-day detention period.

### *Starting a Digester (Slurried Culture)*

Methods of establishing a digester culture differ according to the size of the intended operation.

*Small Digester.* To start a small digester (less than one or two cubic meters), the digester is loaded with the waste to be digested and a "starter." The starter may consist of 10 to 20 lb of soil rich in decayed organic matter; or if available, 40 to 50 gal of bottom mud from a stagnant body of water or swampland. The ideal "starter" would be sludge from a digester already in successful operation. If such a sludge is available, it should be diluted to about 5% solids and enough should be added to constitute about 10% of the full volume of the digester. In the absence of the aforenamed "starters," reliance must be had on time, because a culture must be developed from organisms indigenous to the waste.

*Large-Scale Digester.* A large-scale digester is started in much the same way as is a small-scale one. A difference comes in when access can be had to sewage. If sewage is available, the digester is filled to capacity with sewage, and after a week or so, feeding is initiated and is continued on a reduced scale for 30 days — or until gas production indicates the development of a full complement of methanogens. If digesting sludge is available, the digester is loaded with the sludge to about 10% of the final volume and then is filled with sewage. Loading is initiated immediately, but is based on the volume of the "starter" sludge and not on the total digester volume. For example, if the volume of the "starter" sludge is 350 $ft^3$ and the planned loading rate is 0.12 $lb/ft^3/day$ and the detention period 20 days, then on day 1 the loading would be 44 lb of volatile solids contained in 17.5 $ft^3$ suspension. On day 2, it would be 4.5 lb x (350 + 17.7 $ft^3$) contained in 18.5 $ft^3$ suspension. Thus, the loading becomes progressively greater each day, both in terms of solids and of volume, until the equivalent of the total volumetric capacity of the digester is reached, i.e., at the end of 20 days.

Generally, no problem is encountered in developing the necessary populations of hydrolyzers and acid formers. Difficulties that do arise are with the development of the methanogens. Because of the differences between the microbial populations with respect to rates of development, care must be taken not to allow the pH of the developing culture to drop to an unfavorable level. If the pH level drops below 6.5, loading should be interrupted. Loading can be resumed as soon as gas production either begins or resumes, as the case may be.

## DESIGN

Unless otherwise specified, the material in this section deals with slurried cultures.

### Background

The three types of digestion systems encountered in general wastewater treatment practice are also applicable to the digestion of solid wastes. The three types are the conventional, the high-rate, and the contact systems. High-rate digestion probably is best confined to large-scale operations. It differs from conventional digestion in that it proceeds in two stages and involves the use of two digesters operated in series, as shown by the diagram in Figure 9.3. The active stage takes place in the first digester of the series. The digester contents in unit 1 are thoroughly agitated and the detention period is relatively brief, i.e., a few days. Effluent from unit 1 is passed into unit 2. The culture is kept quiescent in unit 2. Although some digestion takes place in unit 2, the main function of the unit is to serve as a settling chamber in which the digester contents separate into the layers diagrammed in Figure 9.3. In the contact approach, the microbial and hydraulic detention periods are each involved. The workings of the process strongly resemble those of the activated sludge method of wastewater treatment.

### Design Parameters

Design considerations are (1) amounts of waste to be digested; (2) volume of the slurried waste to be added each day as well as the average time a given load will be in the digester (detention time); (3) amounts of end products (gaseous, liquid, solid) discharged each day and their management; and (4) the system for heating and circulating the water used in elevating and maintaining the temperature of the digester culture.

#### *Volume Requirement*

The volume required of a digester is a function of (1) amount of wastes to be processed each day; (2) the moisture content of the waste; (3) the volatile solids concentration; (4) the loading rate; (5) solids content of the slurry; and (6) detention time.

Where means of determining volatile solids content are available, the minimum volume of the digester can be calculated by dividing the amount of volatile solids to be accommodated each day by the desired loading rate. For example, if the volatile solids (VS) of the waste (e.g., manure) to be disposed each day amounts to 2640 lb and the desired loading rate is 0.19 lb VS/$ft^3$/day, then the minimum digester size would be 2640/0.19 or about 13,900 $ft^3$. It should be emphasized that this is the minimum size. Allowances must be made for "freeboard," as well as for adjustments required to reconcile dilution requirements with necessary detention times.

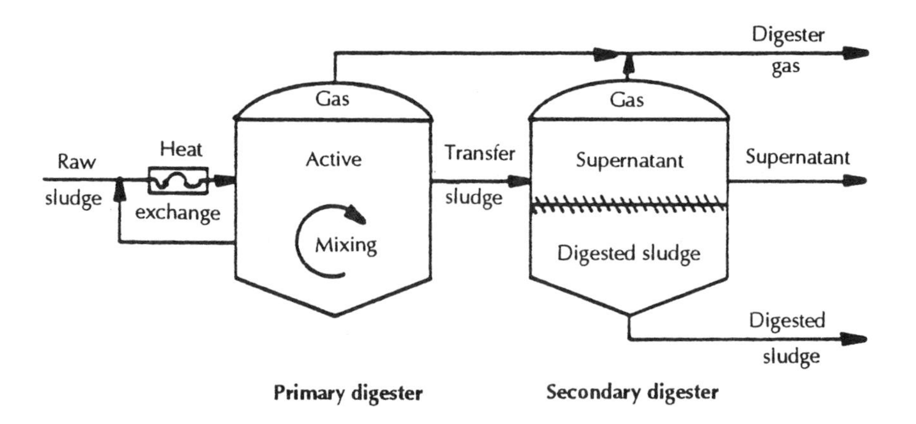


**Figure 9.3.** Two-stage digestion.

The volume required of the gas holder is a function of amount of waste processed per day multiplied by the amount of gas produced per unit mass of waste introduced into the digester. The latter is in turn, a function of the many factors mentioned in the preceding paragraphs. If the gas is to be used on a regular basis, the volume of the holder need be much smaller than the actual volume of gas produced — perhaps 50%.

### *Heat Requirement*

In temperate and cold climates, some heating is necessary to maintain the digestion process at a practical level of activity. Perhaps the easiest way to heat the digester contents is by circulating heated water through a coil immersed in the digester culture. With a large-scale digester, the greatest expenditure of heat energy is that required to elevate the temperature of the incoming feed to the level required for the operation of the culture. The heat needed to warm the input feed is proportional to the mass flow rate and the difference between the feed stream and the temperature of the digester contents. The relation may be expressed as

$$HI = SC(T_1 - T_o)$$

in which S is the feedstream (lb/hr); C, the specific heat content (Btu/lb-°F); $T_1$, temperature of the culture; $T_o$, feedstream temperature (°F); and HI, heat required (Btu/hr). Other heat losses are through convection and radiation to the ambient atmosphere and through evaporation of water vapor from the gas stream. With a large-scale operation, the energy involved in convection and radiation heat losses are minor in comparison to that required to heat the feedstream. What little there is can be compensated by insulating the digester. As the digester volume diminishes, the magnitude of the two losses becomes greater. In that case, heat loss can be minimized by insulating the digester. Although with a sunken digester unit the surrounding soil acts as an insulator, the protection is only at the level of the soil temperature.

Of course, if heat exchange from water circulated in a heating coil is the means of heating (or of any other mode of heating), energy must be expended to elevate the temperature of the water (or other heat-exchange medium). Solar energy is an excellent source of heat energy. An example of the various systems of heating water through solar energy is as follows. The installation is a version of the solar stills used by early researchers to convert saline waters into distilled water. It is essentially a panel with a black backing over which water is trickled. As the water trickles down the inclined panel, it becomes increasingly warmer. The heated water collects in a trough at the base of the panel, from which it is circulated through the hot water coils in the digester. The panel is installed such that it faces the sun. Although the amount of solar energy collected per unit area of collector surface is a function of a number of variable factors, a very rough average estimate for the temperate zones is about 30 kWh/ft$^2$.

## Small Digesters (<35 to about 350 ft$^3$)

### *Examples*

This part of the discussion is focused on the design of relatively small digesters, i.e., those within a range of less than 35 to several cu ft in volume. The discussion concentrates on small units, because safety demands that units larger than a few cubic feet be constructed according to a carefully developed engineering design and of durable materials. Abundant information on designs appropriate for the large units is available in the sanitary engineering literature.

The following are three of the several designs for small digesters given in the literature: The first of the three is of one of several designs developed and tried at the Gobar Gas Research Station in India.[2] Figure 9.4 is a diagrammatic sketch of the design. As the figure indicates, the greater part of the digester is below ground level and is designed to provide heat for the digester culture by means of an immersed hot water coil. The unit can be designed to produce as much as 300 ft$^3$ of gas per day.

The second design was reported by Gotaas in 1956.[15] Sketches of the design are presented in Figures 9.5 and 9.6. Figure 9.5 shows an arrangement of a manure digester connected to two latrines. A detail of the digester is shown in Figure 9.6. Digesters based on the design were in operation on farms in France and Germany in 1955 and 1956. An interesting feature of the design is the direct connection between latrine and digester. The design called for a gas collector unit separated from the digester.

The third design is one developed for use in mainland China,[5,19] and is sketched in Figure 9.7. It should be noted that the gas is stored in the plenum above the culture. The fixed cover of the digester results in an increasing pressure being exerted on the gas as gas production continues. The extent of the pressure depends upon the rate of gas production combined with rate of the usage of the gas.

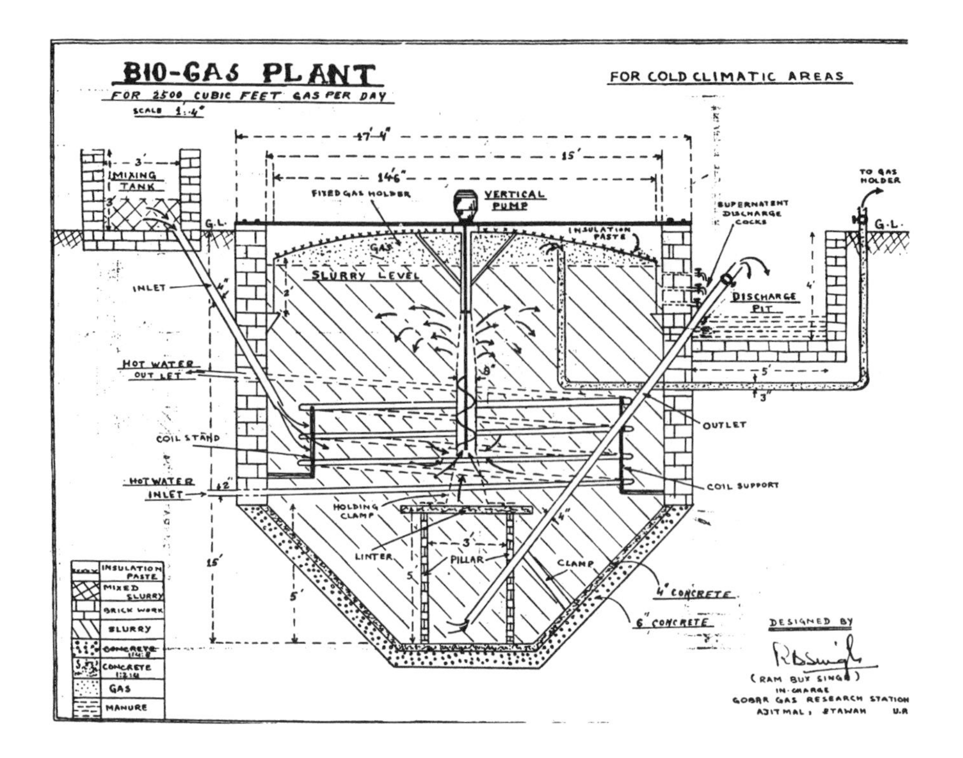


**Figure 9.4.** Diagram of Indian design.

A failing common to the designs shown in Figures 9.5 and 9.7 is the absence of a mixing or agitation system. In the Gobar digester (Figure 9.4), mixing apparently is accomplished by recirculating the digester culture. In addition, as efforts to disseminate the technology increased, workmanship and quality control often decreased. Thus, many units failed due to structural problems and leakage.

### *Design Modifications*

A variation of the preceding designs is the modification of other types of containers to serve as digesters. For example, steel tanks fabricated for water or other storage can be modified to serve as a digester. It is essential that the modification include coating the interior of the tank with a material that is resistant to corrosion by the organic acids formed in digestion. Two or three such installations have been proposed for digesting manure in small-scale (140 or 170 $ft^3$) operations on farms in the U.S.[20]

A design modification based on the use of inexpensive materials indigenous to the site of a proposed operation essentially involves lining an excavated pit with a wall of native materials. The interior of the walled pit is then lined with an impervious material such as plastic film. Suitable provision must be made for capping the pit and collecting and storing the gas. A serious problem with

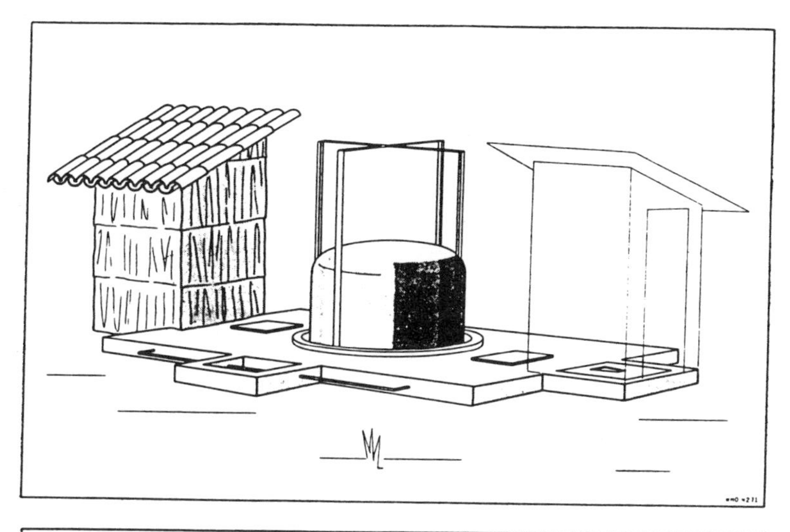

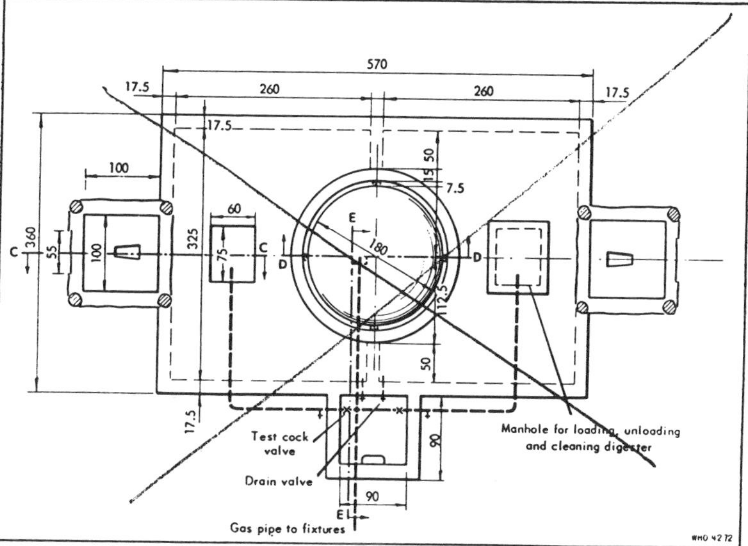


*All measurements are in centimetres.*

**Figure 9.5.** Diagram of manure digester connected to latrines.

such simplified designs is the availability of a suitable lining material. Although plastic film may be impermeable, its durability and degree and permanence of impermeability are open to question. Certain of the organic acids do slowly affect the mechanical properties of plastics.

An example of the simplified approach is the 65-cow dairy plug flow reactor diagrammed in Figure 9.8 and described in Reference 20. Performance of the reactor reportedly compared very favorably with that of a completely mixed reactor.

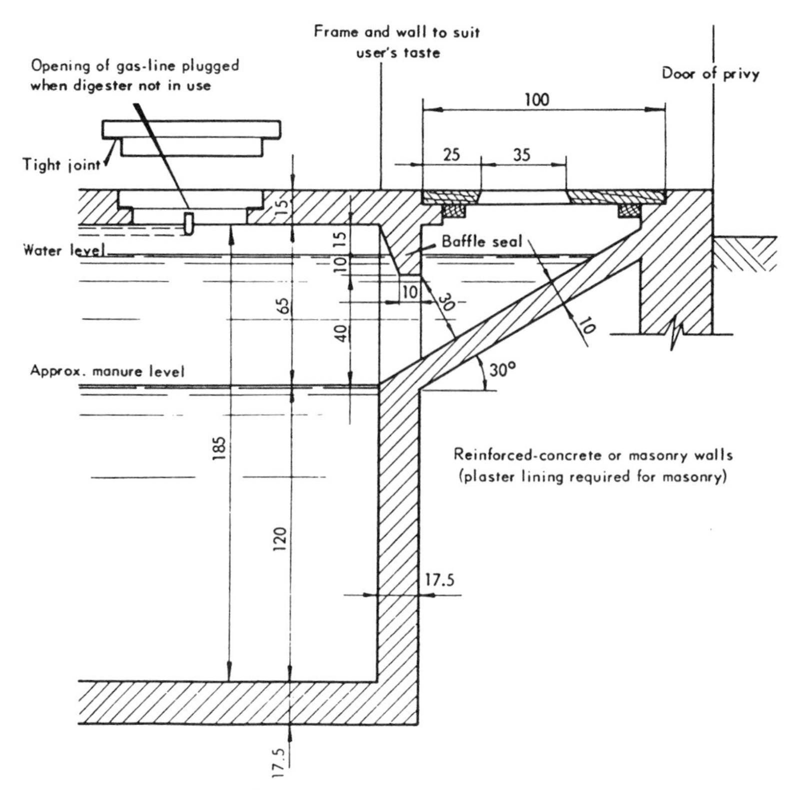


**Figure 9.6.** Detail of manure/night soil digester.

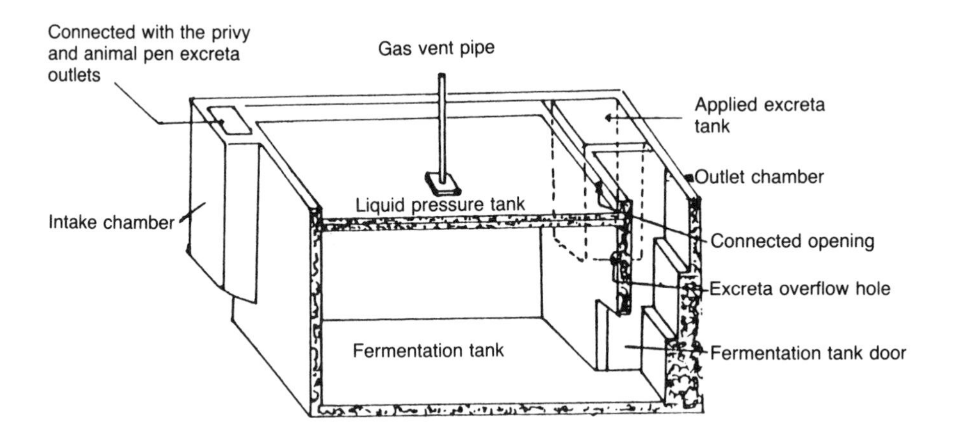


**Figure 9.7.** Chinese design.

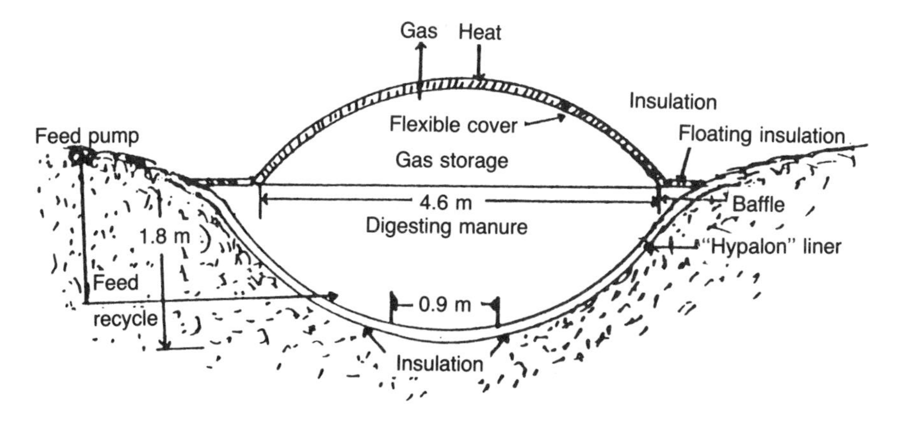


**Figure 9.8.** Plug-flow reactor.

### *Materials*

Type and quantity of materials vary with the size and nature of the installation. An example of materials is a list compiled by Singh.[2] The list is of the materials required for a digester capable of producing about 100 $ft^3$ of gas per day. The list is presented in Table 9.5. From the list, it is possible to arrive at an approximation of types and quantities of materials required for digesters of the design proposed by Singh.

## END PRODUCTS

### Biogas

#### *Composition*

As was indicated in the section on parameters, the major component of biogas is methane (55 to 65%) and the second highest is carbon dioxide (34 to 44%). The remaining gases include $H_2S$, $N_2$, and $H_2O$. The gas as produced (raw gas) has a heat value of 500 to 700 Btu/$ft^3$. It can be used as a fuel for heating purposes or for internal combustion engines. The $H_2S$ can cause a considerable amount of corrosion in internal combustion engines.

Physical and chemical properties that are of especial interest in the utilization of the methane in biogas are listed in Table 9.6. An indication of the amounts of biogas that would have to be produced in a small-scale installation to supply the energy for certain applications is given by the list in Table 9.7.

#### *Upgrading Biogas*

Biogas destined for uses other than simple space heating and household cooking must be upgraded. The upgrading is made necessary by relatively low quality

**TABLE 9.5 Materials Required for A Digester to Produce 2.8 m³ Gas per Day[3,34]**

| Material | Quantity |
|---|---|
| Cement | 40 bags |
| Sand | 8.5 $m^3$ |
| Brick ballast | 2.84 $m^3$ |
| Bricks | 7500 |
| 12- or 14-gauge M.S. sheet drum, 5 ft (1.5 m) in diameter and 4 ft (1.2 m) in height; open at bottom; M.S. angle iron for structure and gas holder guide | 30 m |
| Alkaline pipe 0.5 in. (0.125 cm) in diameter | 15 m |
| Alkaline pipe fittings B end, elbow and sockets of 1 in. (2.5 cm) and 0.5 in. (1.25 cm) fittings | 3 each |
| Wire gauge, 80 mesh | 0.093 $m^2$ |
| Miscellaneous fittings | |
| Paint (enamel) | 3.8 L |

**TABLE 9.6 Physical and Chemical Properties of Methane**

| | |
|---|---|
| Chemical formula | $CH_4$ |
| Molecular weight | 16.042 |
| Boiling point at 14.696 psia (760 mm) | −161.49°C |
| Freezing point at 14.696 psia (760 mm) | −182.48°C |
| Critical pressure | 47.363 kg/cm² |
| Critical temperature | −82.5°C |
| Specifiic gravity | |
| Liquid: −164°C | 0.415 |
| Gas: 25°C and 770 mm | 0.000658 |
| Specific volume: 15.5°C and 760 mm | 1.47 L/g |
| Calorific value: 15.5°C and 760 mm | 38,130.71 kJ/m³ |
| Air required for combustion m³/m³ | 0.27 m³ |
| Flammability limits | 5 to 15% by volume |
| Octane rating | 130 |
| Ignition temperature | 650°C |
| Combustion equation | $CH_4 + 20_2 \rightarrow CO_2 + 2H_2O$ |
| $O_2/CH_4$ for complete combustion | 3.98 by weight |
| $O_2/CH_4$ for complete combustion | 2.0 by volume |
| $CO_2/CH_4$ from complete combustion | 2.74 by weight |
| $CO_2/CH_4$ from complete combustion | 1.00 by volume |

of raw biogas. Moreover, its day-to-day composition is quite different from that of natural gas. An undesirable difference is the low and varying heat content of biogas. The variation may be from 200 to 600 Btu/ft³. The minimum heating value of natural gas is 1000 Btu/ft³. Another undesirable feature is the moisture content of biogas. It may be as low as 5% or as high as saturation. If the biogas is produced in a landfill, its $O_2$ content could be sufficiently high to render the gas explosive. The variation of the preceding parameters from day to day and season to season may be as much as 50%. Other undesirable characteristics are a sizeable $CO_2$ and $N_2$ contents. These characteristics lower the heating value of the biogas.

Biogas can be upgraded through one or more of several procedures, all of which are relatively expensive and require special equipment. The procedures may be as simple as dehydration or as extensive as $H_2O$, $CO_2$, and $N_2$ removal. Dehydration can raise the heating value by about 10% of the original value. The

**TABLE 9.7 Quantities of Biogas Required for Various Uses[a]**

| Use | Specification | Quantity of Gas Required ($m^3$/hr) |
|---|---|---|
| Cooking | 5 cm burner | 0.33 |
| | 10 cm burner | 0.47 |
| | 15 cm burner | 0.64 |
| | 5–10 cm burner | 0.23–0.45 |
| | per person/day | 0.34–0.42+ |
| Gas lighting | per mantle | 0.07 |
| | per mantle | 0.07–0.08 |
| | 2-mantle lamp | 0.14 |
| | 3-mantle lamp | 0.17 |
| Gasoline or diesel engine[b] | converted to biogas, per hp | 0.45–0.51 |
| Refrigerator | per $m^3$ capacity | 0.98 |
| | per $m^3$ capacity | 1.20 |
| Incubator | per $m^3$ capacity | 0.5–0.6 |
| | per $m^3$ capacity | 0.5–0.7 |
| Gasoline | 1 L | 1.33–1.87[c] |
| Diesel fuel | 1 L | 1.50–2.07[c] |
| Boiling water | 1 L | 0.11[d] |

[a] Adapted from References 3 and 34.
[b] Based on 25% efficiency.
[c] Absolute volume of biogas needed to provide energy equivalent of 1 L of fuel.
[d] Absolute volume of biogas needed to boil off 1 L of water.

heating value is brought up to 600 to 700 Btu/ft$^3$ through dehydration accompanied by carbon dioxide and hydrogen sulfide removal.

Dehydration methods include use of in-line gravity outflow, filtering, special solvents (polyethylene glycol), molecular sieves, heating, air cooling, and refrigerant cooling. Of the several methods, the use of molecular sieves is one that is both relatively inexpensive and yet has a high degree of efficiency. Because of the highly localized polar charges characteristic of molecular sieves, polar or polarizable compounds are strongly adsorbed on the molecular sieves. Moreover, the absorptive capacity is much greater than that of other absorbents.

Because of the unusual hygroscopicity of glycols, their excellent thermal and chemical stability, low vapor pressures, and ready availability at moderate cost, the triethylene glycol (TEG) system is widely used for gas dehydration. As the gas enters the system, it is compressed, and bulk contaminants are removed in a knockout drum. After compression and cooling to remove the bulk of the water, the gas enters a TEG adsorber/separator tower. Free liquids in the gas stream are removed in the lower part of the tower (the separator section) as the stream ascends to the upper or absorber section of the tower. In the absorber section, the gas stream contacts lean TEG on bubble-cap trays.

By coupling the TEG dehydration system with a hot potassium carbonate scrubbing system, dehydration and carbon dioxide and hydrogen sulfide removal can be accomplished simultaneously.

If only the hydrogen sulfide need be removed, it can be readily done by passing the gas through a dry gas scrubber containing an "iron sponge" made up of ferric oxide mixed with wood shavings; 1 bu (1.2 ft$^3$) of iron sponge is

**TABLE 9.8 Disease-Causing Organisms Found in Human Excrement**

| Category | Disease | Organisms (where identified) |
|---|---|---|
| Viral | Infectious hepatitis | |
| | Gastroenteritis | |
| | Respiratory illness | Adenovirus |
| | | Reovirus |
| | Poliomyelitis | Enterovirus (poliovirus) |
| Bacterial | Typhoid fever | *Salmonella typhosa* |
| | Salmonellosis | *Salmonella* spp. (Exp. *S. paratyphi, S. schottmueleri*) |
| | Bacillary dysentery (Shiegellosis) | *Shigella* spp. |
| | Cholera | *Vibrio cholerae* |
| | Tuberculosis | *Mycobacterium tuberculosis* |
| Protozoan | Amebiasis (amoebic dysentary) | *Entamoeba histolytica* |
| Helminthic | Roundworm | *Ascaris lumbricoides* |
| | Pinworm | *Oxyaris vermicularis* |
| | Whipworm | *Trichurus trichiura* |
| | Tapeworm | *Taenia saginate* |
| | Hookworm | *Ancylostoma duodenale* |
| | | *Necator americanus* |

sufficient for removing about 8 lb of sulfur. The ''sponge'' is regenerated by exposure to air. Ferric sulfide formed in the scrubbing operation is converted into ferric oxide and elemental sulfur.

Upgraded gas can be used on-site to generate electricity, or it may be injected into a public utility transmission line. With on-site electrical generation, the gas can be fed to an internal combustion engine or burned to drive a gas turbine. The gas should be compressed to about 5 psig for use in an internal combustion engine (diesel engine). If a gas turbine is used, the pressure must be increased to 150 psig.

*Costs.* Although costs of upgrading biogas vary not only from installation to installation, but also from country to country, and because of the inflation factor, a few overall costs are given to provide a basis for judging economic feasibility. In 1990 the cost was about $2.30 per Btu per kJ to dehydrate and compress (420 psig) raw gas at 1 million ft$^3$/day in the U.S. Gas thus treated had a heating value of 450 Btu/ft$^3$ and a moisture content of 7%. At the same daily throughput, the cost to dehydrate the raw gas to 7% moisture, remove its carbon dioxide and hydrogen sulfide, and compress it to 420 psig was $3.80 per million Btu. The gas would have a heating value of 600 to 700 Btu/ft$^3$. A stream factor of 0.85, 10-year landfill life span, 20-year equipment life, and a 12% rate of return are assumed. Noteworthy is the fact that compression would account for at least 50% of the costs.

*Comment.* Because of the costs and complexity of the equipment and operation involved, upgrading biogas to pipeline quality would not be recommended for application in developing countries. The exception would be one in which the wastes from a highly developed metropolitan area were disposed in a sanitary landfill. The practical procedure in almost all cases would be to burn the gas directly at the site of generation and to put the heat energy to some immediate use.

## Residues

The principal residues of biogasification that require management and treatment are the supernatant and the settleable solids. The supernatant, as stated earlier, refers to the primarily liquid portion of a quiescent digester culture, i.e., the layer between the scum and the sludge layers (cf. Figure 9.2). The solids or sludge layer is the bottom-most one in the digester. In common parlance, the term "sludge" is applied indiscriminately to the solids in the sludge layer and to the combined solids and supernatant. An alternative approach is to subdivide the solids layer into two layers, namely, the digesting sludge and the inactive sludge layers. For the purposes of this book, the term "sludge" is restricted to the sludge layer, which in turn is treated as a single layer.

### *Supernatant*

The supernatant is a very concentrated suspension of solid particles of colloidal dimensions, of bacterial cells, and of dissolved solids. Nonliving material in the supernatant is biologically highly unstable. Consequently, untreated supernatant cannot be disposed of in the surrounding environment without incurring a seriously unfavorable impact.

In an ongoing operation, a substantial portion of the supernatant is recirculated into the digester. In a two-stage operation, the supernatant is returned to the first digester in the series. Recirculating into the digester system is very useful because of the magnitude of the microbial populations suspended in the supernatant and of the appreciable concentration of unutilized nutrients. Unrecirculated supernatant can be spread upon the land along with the sludge.

### *Sludge*

The solids content of the sludge immediately after it is discharged from a slurry-type digester is rather low — generally within a range of about 5 to 10%. It may be applied to the soil directly or it may be further dewatered. As far as developing nations are concerned, the most suitable method of dewatering is through drainage and evaporation on a sand bed. A properly digested sludge spread upon a well-drained sand bed dewaters to 15 to 20% solids within a week or two of sunny, dry days. During the rainy season the beds should be sheltered from the rain.

The nature of the sludge is largely influenced by that of the waste introduced into the digester. In general, the sludge externally resembles the product that would be formed if the waste were aerobically composted. Sludge usually bears a strong external resemblance to an aerobically composted product. Except for their slightly higher nitrogen content, sludges also resemble compost with respect to chemical characteristics. However, it is advisable to compost sludge before applying it to the land, especially if human excrement had been included in the raw waste. The upshot of these resemblances is that the sludge can be used in agriculture in much the same manner as would be composted material. However, certain constraints of a public health nature apply to the use of sludge.

## LIMITATIONS

Limitations on biogasification are of four major types, namely public health, technical, socio-economic, and organizational.

### Public Health

Public health problems arise mainly when human excrement are a part of the feedstock. As used in this discussion, the term ''excrement'' includes collected fresh excrement, night soil, septic tank cleanings, raw sewage sludge, and any other material that may contain human excrement. The hazardous nature of human excrement and of wastes containing excrement is due to the likely presence of disease-causing (pathogenic) organisms in the excrement. Numbers of such organisms present in excrement are especially great in the warmer regions of the world, and where sanitation is inadequate. The list of pathogenic organisms in Table 9.8 is an indication of the types.

The danger from disease causing organisms is greatest during the handling of the excrement or excrement-containing wastes and in the consumption of crops grown on and on which excrement had been spread. The number of individuals endangered during handling is much less than that of those affected as a result of spreading the material on the land. The handling hazard can be minimized through the imposition of sanitation measures, which would include careful personal hygiene on the part of the workers.

It should be pointed out that human excrement is not the sole source of disease-causing organisms in wastes to be digested, although it is by far the major source. Animal waste and animal products may contain microorganisms pathogenic to man, i.e., zoonotic organisms. Examples are leptospira, parasites, and anthrax spores.

Potential public health hazards arising from the use of digester residue in agriculture are twofold: (1) those from the presence of disease-causing organisms; and (2) those from the presence of heavy metals and toxic chemicals. As stated in the preceding paragraphs, disease-causing organisms become a concern only when human excrement is one of the wastes introduced into the digester. Although the destruction of pathogens in the biogasification process is substantial, it is not complete, as is indicated by the data in Table 9.9. Extent of destruction increases with temperature and prolongation of the detention period. Because destruction is not complete, utilization of the sludge in the production of root crops to be eaten uncooked or unpeeled should be minimal. Studies have shown

**TABLE 9.9 Die-Off of Enteric Disease-Causing Organisms During Anaerobic Digestion**

| Organisms | Temperature (°C) | Residence Time (days)[a] | Die-Off (%) | Ref. |
|---|---|---|---|---|
| Poliovirus | 35 | 2 | 98.5 | 20 |
| *Salmonella* ssp. | 22–37 | 6–20 | 82–96 | 35 |
| *Salmonella typh.* sp. | 22–37 | 6 | 99 | 35 |
| *Mycobacterium tuberculosis* | 30 | | 100 | |
| *Ascaris* | 29 | 15 | 90 | 35 |
| Parasite cysts | 30 | 10 | 100[b] | 36 |

[a] Time in digester.
[b] Does not include *Ascaris.*

that die-off of polio virus in digester maintained at mesophilic temperature can be at least 98% *Salmonella sp.* and *S. typhosa* 82 to 92%; of parasite cysts (exclusive of *Ascaris*), 99 to 100%; and of *Ascaris*, 90%.[21] Tabulated data on die-off are given in Table 9.9.

The sludge can be rendered safe in terms of pathogen kill if it is subject to composting. Consequently, composting is advisable in an operation in which human excrement is a raw material.

As a rule, heavy metals and toxic chemicals become a potential problem only when industrial wastes are involved. Biogasification has no effect on heavy metals and on most toxic chemicals. It does reduce the concentration of some halogenated hydrocarbons. The increasing stringency of regulations regarding disposal of heavy metals and toxic chemical into the municipal sewerage system has substantially reduced the concentrations of those contaminants in municipal wastewaters and, consequently, in the sludge. Problems resulting from the presence of heavy metals as well as methods of alleviating them are discussed in the chapter on the use of organic wastes as a fertilizer.

## Technical Limitations

As stated earlier, biogasification has the limitations characteristic of biological processes. Chief among them is the inability of complexity and sophistication of equipment and optimization of conditions to increase the overall capacity and efficiency of the process beyond that permitted by the genetic makeup of the microorganisms involved.

In practice, technical limitations are more in the nature of constraints, problems, and diminishing returns with respect to efficiency, rather than of upper physical boundaries. This is especially true with respect to percent energy recovery. For example, not all of the energy bound in the biodegradable organic fraction of MSW and of many agricultural wastes can be recovered through biogasification because not all of the reaction is readily available to the microorganisms. It is true that the availability can be increased by subjecting the resistant fraction to physical and chemical treatment, but doing so is expensive in terms of energy and monetary outlay, i.e., beyond the point of diminishing returns.

When making an energy balance for biogasification, it is important that an account be made of each and every expenditure of energy involved in carrying out the process. This accounting applies to the energy involved in collecting material that otherwise would not reach the operation, and to the energy expended in preparing the waste, in the conduct of the process, and in the preparation of the biogas for use. (Strictly speaking, the accounting would include the energy expended in the manufacture, etc. of the equipment.)

Diminishing return is an important factor with respect to the size of the reactor. It arises from the need to make the reactor and gas storage equipment air tight, so as to avoid inhibition particularly of the methane formers and of fire and explosion. Maintenance of air tightness becomes more difficult with increase in size of the units involved. Overall management also becomes more intricate and hard to do as size becomes larger. A solution to the size difficulty would be the modular approach.

With MSW and certain industrial and agricultural wastes, feedstock preparation may present technical difficulties; although the difficulties do not necessarily increase with size. Feedstock preparation is much the same as that required for composting. However, it is more of a problem with slurry-type digester systems than with high-solids digestion.

The difference comes from the need to slurry wastes not already in that form and to contain the slurry. Slurrying involves two main steps: (1) reduction of particle sizes of incoming material to a point at which the material can be slurried (maximum size 0.5 or 1 in.); and (2) suspending the particles in an aqueous medium (usually water). Another operational problem peculiar to slurry-type biogasification is the accomplishment of adequate mixing of the digester contents. Adequate mixing becomes increasingly difficult with increase in size of the reactor. A final difference is the need to dewater the residue slurry. Unless the digested slurry is dewatered by gravity filtration (sand bed), size of operation makes little difference. Surface area requirements would be serious when large volumes of slurry are involved.

Although the high-solids type of biogasification may be free of slurrying problems, preparation of its feedstock and the attending problems are comparable to those for composting.

Among other major technical problems are corrosion, particularly of gas holders, wear and tear of the various components, deterioration of the reactor waller and hoses, and cracking of the digester walls.

Although biogasification would seem to be an excellent means of managing and treating farm wastes, particularly manure, it has some drawbacks which lead to a negligible number of farm-scale operations. Among the major reasons for the low number are (1) the monetary and spatial expenditures involved in installation and operation; (2) the addition of another task be performed by the already overburdened farmer; and most of all, (3) the low cost of fossil fuel and available energy; and (4) the relatively inferior value of raw biogas as a fuel (energy source) as compared to available fossil fuel. In terms of cost, upgrading the biogas would make it even less competitive with natural gas. In temperate climates, the disadvantages are amplified by the decline in treatment performance and gas production during cold seasons.

## SUMMARY

Although the scientific aspects of biogasification are fairly well understood and low-cost designs have been developed and proposed, biogasification has a long way to go before it is widely regarded as being competitive with composting. The distance is even greater with respect to MSW. At this time, the status of biogasification might be said to be at the level characteristic of composting in the 1970s. Despite claims to the contrary, practical experience with biogasification has been very limited. The problem is that sufficient practical experience is a necessary prelude to significant progress. The future is more bright because energy inevitably will become less abundant and consequently become more expensive. Although the percentage energy recovery from wastes may not equal that with incineration, the environmental and cost problems are far fewer and much less intense.

## RECOVERY OF BIOGAS FROM LANDFILLS

### Introduction

The search for energy sources has led to the tapping and subsequent upgrading of the biogas generated through decomposition of the wastes buried in landfills. It has long been known that pockets of biogas begin to form in landfilled waste within a few months after a fill has been established. In fact, biogas generation, particularly the methane fraction, has long been regarded as one of the problems associated with landfills, especially sanitary landfills. Biogas is less likely to escape from a sanitary landfill because enclosure of the decomposing waste is more complete than in simple landfills. The origin of the methane is the anaerobic fermentation, (i.e., biogasification) of the organic fraction of the buried wastes.

Biological principles involved in biogasification are identical with those described under Principles. In a landfill, differences from conventional biogasification are more in the nature of environmental responses than in the process itself. For example, unlike conventional biogasification, the wastes are not subjected to preparatory steps designed to enhance digestibility, nor is inorganic material removed. No serious attempt is made to optimize environmental conditions. The significance of these differences is that in a landfill, the rate of biogasification is slower and the size of the fraction of the organic mass eventually converted into methane is less than in conventional biogasification. Consequently, methods for estimating gas production in a conventional digester are only partially suitable for predicting methane production in a landfill. Generally, the amount of methane actually obtained from a landfill is much less than the theoretical volume predicted with the use of existing methods and models. The specific inclusion of gas containment and eventual collection in the design of a landfill will result in an increase in yield of gas.

## Volume and Composition of Biogas

The initial biodegradation, i.e., directly after the wastes are buried, is aerobic to a certain extent because of the oxygen contained in the air entrapped in and with the buried wastes. The duration of this stage may be as brief as a few days or as long as several weeks. Because the compaction and covering of the wastes excludes the replenishment of the entrapped air, the oxygen is gradually depleted and eventually biodegradation becomes anaerobic.

The transition from aerobic to anaerobic decomposition and the attendant methane production of the latter proceeds as a series of phases. The first phase is the aerobic phase. Its duration is the time required to use up the entrapped oxygen. This may be days or weeks. The predominant gas generated during this stage is carbon dioxide. The second phase is the transition from aerobiosis to anaerobiosis. It is characterized by the die off of obligate aerobes and a shift to the anaerobic mode by facultative aerobes. The principal gases produced during transition are carbon dioxide ($CO_2$) and hydrogen ($H_2$). The latter is produced to a lesser extent. The third phase is marked by the gradual appearance of methane ($CH_4$). In the fourth (final) phase, $CH_4$ production becomes constant. The composition of the gases during this phase (anaerobic phase) is on the order of 40 to 50% $CH_4$, 30 to 40% $CO_2$, 10 to 20% $N_2$, 1% $O_2$, and traces of sulfides and volatilized organic acids. The data in Table 9.10 indicate the composition of a gas produced in a typical landfill. As is indicated by the data in Table 9.11, the gas may also contain traces of volatile organic compounds that may have been disposed with the refuse.

Rate and volume of gas production are functions of wastes disposed and of the conditions prevailing in a landfill. Among the factors that affect rate are temperature, pH, and moisture content. Amount of the buried wastes and age of the landfill are among the factors that affect the volume of gas production. The wide variation of rates and volumes of production from one region to another is due to a corresponding variation in wastes and conditions.[22-27] An example of the variation of rates is as follows: 40 scf gas/ton/year of waste disposed to 240 scf/ton/year; and of volumes, 1 to 7 $ft^3$/lb of refuse disposed. Although gas production may continue in gradually dwindling amounts as long as 50 years, the greatest part of it is accomplished within about 20 years after closure and is most active during the first 5 years.

Among the several models developed to predict rates and volumes of gas from landfills is the relatively rigorous stoichiometric approach (i.e., relative to other approaches) described in Reference 22. In the model, biodegradable waste converted into biogas is divided into two major classes based on ease of degradability. The classes are (1) the easily biodegradable fraction (e.g., food waste or garbage, garden debris); and (2) the less easily biodegradable fraction (e.g., paper, textiles, etc.). However, this model and others developed to predict rates and volumes of biogas should be regarded only as approximate indicators of expected gas production trends. Most models require actual measurements of gas production

**TABLE 9.10 Typical Composition of Landfill Gas**[37]

| Component | Component Percentage (dry volume basis) |
|---|---|
| Methane | 47.5 |
| Carbon dioxide | 47.0 |
| Nitrogen | 3.7 |
| Oxygen | 0.8 |
| Paraffin hydrocarbons | 0.1 |
| Aromatic and cyclic hydrocarbons | 0.2 |
| Hydrogen | 0.1 |
| Hydrogen sulphide | 0.01 |
| Carbon monoxide | 0.1 |
| Trace compounds[a] | 0.5 |

[a] Trace compounds include sulfur dioxide, benzene, toluene, methylene chloride, perchloroethylene, and carbonyl sulfide in concentrations up to 50 ppm.

**TABLE 9.11 Trace Organic Compounds in Raw Landfill Gas Mountain View Landfill, 1980**[38]

| Compound | Concentration ($mg/m^3$) |
|---|---|
| 1,2-Dichloroethylene | 5.2 |
| Trichloroethylene | 10.4 |
| Methyl isobutyl ketone | 5.1 |
| Chlorobenzene | 0.4 |
| Toluene | 4.0 |
| Tetrachloroethylene | 4.5 |
| Ethylbenzene | 4.0 |
| Xylene | 2.3 |

in order to determine the values of constants for the models. The variables mentioned in the preceding paragraphs together with others make such an attitude appropriate.

Among the variables that especially affect the accuracy of prediction models are these two: (1) volume of gas that escapes the fill; and (2) the percentage of carbon in the landfilled wastes that passes through the methane fermentation route and that becomes a part of microbial protoplasm.

## Collection of Biogas

If biogas recovery is planned for a new landfill, certain features should be incorporated in the design of the fill. Some of the features are characteristic of modern landfill design — particularly those features related to effectively sealing off the landfilled waste from the land, water, and air environments.

Features specific to biogas recovery are those used in controlling the movement of gas to facilitate collection. This is done by installing a combination of strategically spaced wells and areas of high permeability through which the gas is channeled to collection points. Doing this involves the installation of underground venting pipes and a gravel layer between a liner and the waste, or gravel filled trenches. The gas is removed (i.e., extracted) from the landfill by way of a piping or header system to transport the gas and a blower to withdraw the gas from the fill through the headers,[22, 23, 28] as shown in Figure 9.9.

Proper functioning of the gas collection system is ensured through the operation of blowers to produce a partial vacuum in the headers and collection system and thereby pull the biogas from the landfill. This is done despite the fact that some gas would flow unassisted into the collection wells because of the slightly elevated internal pressure of the landfill. However, the flow rate would be too low to ensure proper collection performance. Moreover, blowers both increase the flow of gas from the landfill and broaden the effective landfill area serviced by each gas well.

The blowers can be adjusted either to (1) pull gas from the fill and discharge it at atmospheric pressure for dispersion, flaring, or combustion; or (2) compress the gas to higher pressures for distribution or for further processing.

Gas can be recovered from a landfill not originally designed to accommodate biogas collection. This is done by way of drilling a number of boreholes into the landfill at selected gas collection points. The boreholes should be 2 to 3 ft in diameter. Their depth should be from 50 to 90% of that of the buried refuse. The boreholes are fitted in the same manner as the collection wells used in fills designed for gas recovery.

The collection wells are packed with gravel and are equipped with casings that extend the full depth of the fill. The casings are perforated in the section exposed to the contents of the fill. They must have telescopic connections between pipe segments such that connections between segments are maintained despite significant and nonuniform subsidence of the fill.

The wells are built by progressively backfilling gravel around the gas collection pipe. The backfilled gravel (or a coarse substitute) serves as a highly permeable collection zone through which the gas flows into the collection pipe for removal from the well. The gravel area is covered with a gas-tight seal topped by backfilled soil to form a barrier against intrusion of external air into the well.

Air intrusion into a well (or into any part of the fill) dilutes the collected gas and thereby lowers its heating value. The air also complicates the upgrading of the biogas. With respect to dilution, the concentration of nitrogen in the collected gas is increased, thereby lowering the quality of the gas. The oxygen content of the intruding air inhibits the activity of the methane-forming microorganisms and may raise the methane to oxygen ratio to an explosive level.

The arrangement of the collection wells is determined by (1) their respective capacities; (2) the characteristics of the soil cover; and (3) provisions for directing gas movement in the fill. The dimensions of the fill area affected by a well is a function of the rate of pumping. For example, the negative pressure in a fill 40 ft deep and provided with a gas well 20 ft deep ranged from −2 in. of water at the well to less than −0.3 in. at a distance of 100 ft from the well when the

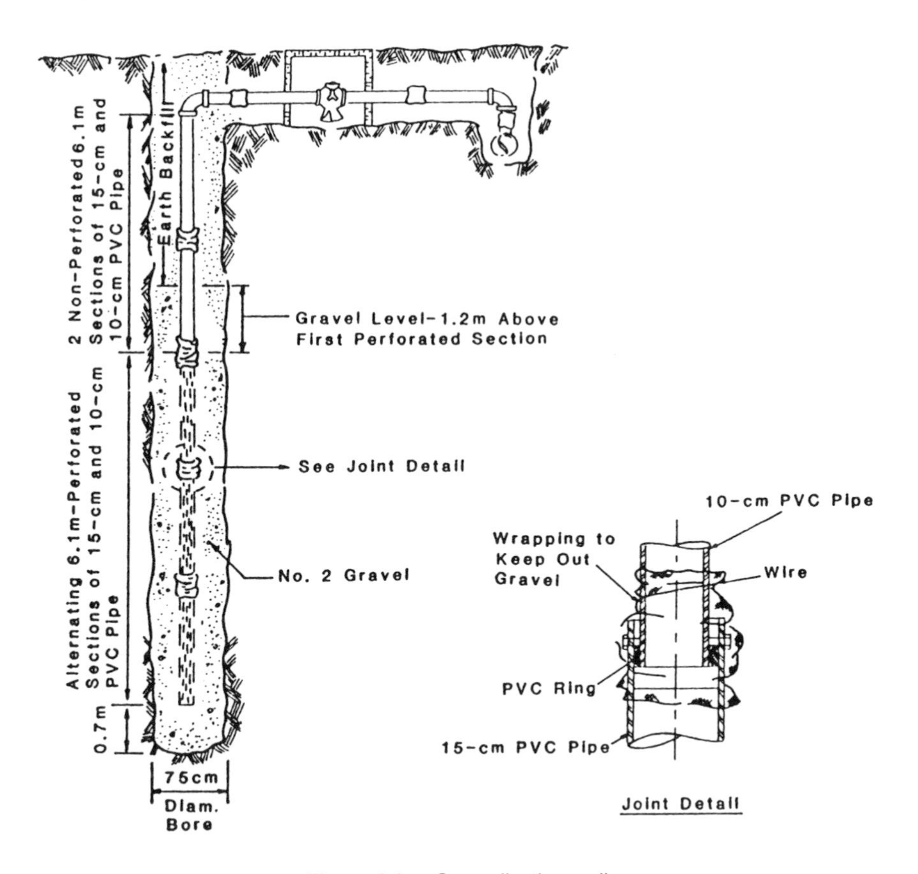


**Figure 9.9.** Gas collection well.

well was being pumped at 100 scf/min. Advancing the pumping rate to 300 scf/min brought the respective negative pressures to −7 and −1 in. of water.

It is important that the gas not be withdrawn at a rate great enough to pull air through the cover and into the fill, especially if the cover material is relatively porous.

## Upgrading and Utilization

Although upgrading and utilization were covered in the section End Products, they also are described in this section, but only as they pertain to biogas collected from landfills. Nevertheless, some repetition is inevitable.

The biogas should be upgraded before being put to use. Even if used for simple space heating and household cooking, at least a modicum of upgrading is indicated. The required upgrading is much more extensive if the gas is to be used as a fuel for an internal combustion engine, or is to be injected into existing transmission lines.

## Quality and Heat Content

As is true of biogas in general, the quality and heat content of landfill gas do not compare favorably with those of natural gas. The quality is further lowered by its variability in composition and other characteristics. Thus, its heat and moisture contents and oxygen concentration may vary as much as 50% from day to day and season to season. As a rule, the heat content of landfill biogas ranges from 200 to 600 Btu/ft$^3$, whereas the lowest heat content of natural gas is 1000 Btu/ft$^3$. The moisture content of landfill gas may range from 5% to saturation. Oxygen ($O_2$) content varies from trace levels to levels that are potentially explosive. However, explosive levels are reached very infrequently. Finally, its usually sizeable carbon dioxide and nitrogen contents materially lower its heat content and, hence, the quality of the gas.

Ranking high among the uses for upgraded landfill biogas are on-site generation of electricity and/or injection into a public utility transmission line. For on-site generation of electricity, the gas is used to fuel an internal combustion engine or to drive a gas turbine. If the gas is to be used in an internal combustion engine, it is compressed to about 5 psig. For a gas turbine, the pressure is increased to 150 psig.

Dehydration (removal of moisture) can raise the heating value of the gas by about 18%. Dehydration accompanied by carbon dioxide and hydrogen sulfide removal results in a heating value of 600 to 700 Btu/ft$^3$. Among the dehydration procedures are (1) inline gravity outflow; (2) filtering; (3) use of special solvents (e.g., glycol, polyethylene); (3) passage through molecular sieves or permaselective membranes; and (4) subjection to heating, air cooling, and refrigerant cooling. Of the four procedures, passage through a molecular sieve combines a relatively low cost with high efficiency.

The TEG system is widely used for dehydrating landfill gas. Reasons for its extensive use are the high degree of hygroscopicity of glycols, their excellent thermal and chemical stability, low vapor pressures, and moderate cost.

Certain uses (e.g., space heating, household cooking) only require that $H_2S$ be removed. Hydrogen sulfide can be removed by passing the gas through a dry-gas scrubber that contains a mixture of ferric oxide and wood shavings (''iron sponge''). The removal capacity of the mixture is about 7 lb of sulfur per cubic foot of mixture. The mixture can be regenerated by exposing it to air. Doing so converts the ferric sulfide formed in the scrubbing operation to ferric oxide and elemental sulfur.

## Economic Feasibility Factors

Several factors exert a decisive influence on the economic feasibility and advisability of recovering gas from a landfill and putting it to use. Among the more important factors are (1) size and location of the fill; (2) permeability of cover material and surrounding soil layer; (3) climatic conditions; and (4) intended use of the gas.

With regard to permeability of surrounding soil layer, it is far more feasible to provide for an impermeable barrier between the landfill contents and the surrounding soil while the fill is as yet in the design stage than to install one after the fill has been completed. If the latest sanitary landfill design criteria are followed, permeability of cover and surrounding soil layer should not be a problem. Nothing much can be done about the size and location of a completed fill or of one presently in use. The same can be said of climatic conditions. If the intended use requires a top quality gas, cost of upgrading and technological infrastructure must be carefully considered.

The mass of waste in the fill should be sufficient to ensure an eventual total gas output that would have a monetary and energy value in excess of that expended on necessary departures from conventional fill practice. The size of the fill must be great enough to ensure gas production over a period sufficiently long to warrant the installation of equipment needed for collecting, upgrading, and using the gas.

It would not be advisable to utilize a fill that is less than 40 ft deep. The completed fill should contain at least about two million tons of municipal solid waste.[2] The peak rate of generation (raw gas production) from such a fill would be from 1000 to 1200 $ft^3$/min.

## REFERENCES

1. Pacey, J. "Landfill Gas Recovery," in *Biogas and Alcohol Production* (Emmaus, PA: J.G. Press, 1980).
2. El-Halwagi, M.M., Ed. *Biogas Technology, Transfer, and Disposal* (New York: Elsevier, 1984).
3. Singh, R.B. *Bio-Gas Plant: Generating Methane from Organic Waste*, Gobar Gas Research Station, Ajitmal, Etawah, India.
4. Blobaum, R. "Biogas Production in China," in *Biogas and Alcohol Production* (Emmaus, PA: J.G. Press, 1980).
5. McGarry, M.G. and Stainforth, J., Eds. *Compost, Fertilizer, and Biogas Production from Human and Farm Wastes in the People's Republic in China*, Industrial Development Research Center, PO Box 8500, Ottawa, Ontario, Canada, K1G 3H9, or 60 Queen St., Ottawa (1978).
6. "Methane Generation from Human, Animal, and Agricultural Wastes," National Research Council, National Academy Sciences, Washington, DC (1977).
7. Babbitt, H.E. *Sewerage and Sewage Treatment*, 6th ed. (New York: John Wiley & Sons, 1947).
8. Babbitt, H.E., B.J. Leland, and F.B. Whitley, Jr. "The Biological Digestion of Garbage with Sewage Sludge," Bull. No. 287, Engineering Experiment Station, University of Illinois, Urbana, November (1936).
9. Fair, G.M., J.C. Geyer, and D.A. Okun. *Elements of Water Supply and Wastewater Disposal*, 2nd ed. (New York: John Wiley & Sons, 1971).

10. Diaz, L.F. ''Energy Recovery through Biogasification of Municipal Solid Wastes and Utilization of Thermal Wastes from an Energy-Urban-Agro-Waste Complex,'' Doctoral Dissertation, University of California, Berkeley (1976).
11. Diaz, L.F. ''Overview of Selected U.S. Methane Recovery Installations,'' in Proc. 1979 Biogas, Alcohol Seminar, Chicago, October 1979.
12. Pfeffer, J.T. and J.C. Liebman. ''Biological Conversion of Organic Refuse to Methane,'' annual report, NSF/RANN/Se/GI-39191/75/2, Dept. C.E., University of Illinois, Urbana report, UILU-ENG-75-2019, September 1975.
13. Jewell, W.J. ''Future Trends in Digester Design,'' in *Proc. of the First Int. Symp. on Anaerobic Digestion*, Cardiff, Wales, Applied Science Publishers Ltd. London.
14. Diaz, L.F., G.M. Savage, G.J. Trezek, and C.G. Golueke. ''Biogasification of Municipal Solid Wastes,'' in *Proc. of the North National ASME Waste Processing Conf.*, May 1980.
15. Gotaas, H.B. *Composting — Sanitary Disposal and Reclamation of Organic Wastes*, WHO Monograph Series No. 31 (1956).
16. Golueke, C.G. ''Composting Manure by Anaerobic Methods,'' *Compost Sci.*, 1(1):44–45 (Spring 1960).
17. Golueke, C.G. *Composting: A Study of the Process and Its Principles* (Emmaus, PA: Rodale Press, 1972).
18. Golueke, C.G. ''Temperature Effects of Anaerobic Digestion on Raw Sewage Sludge,''*Sewage Works and Ind. Wastes*, 30:1142–1225 (October 1958). (*SW&IW* later became the *Journal of the Wastewater Pollution Control Federation.*)
19. Saubolle, B. and A. Bachmann. *Fuel Gas from Cow Dung* (Kathmandu, Nepal: Sahayogi Press, April 1980).
20. Hayes, T.D. et al. ''Methane generation from small scale farms,'' in *Biogas and Alcohol Fuels Production* (Emmaus, PA: J.G. Press, 1980).
21. Golueke, C.G. ''Epidemiological Aspects of Sewage Sludge Handling and Management,'' in Proc. Int. Symp. on Land Application of Sewage Sludge, Tokyo (October 1982).
22. Ham, R.K. et al. ''Recovery, Processing, and Utilization of Gas from Sanitary Landfills,'' EPA-600/2-79-001, U.S. EPA, Cincinnati, OH (1979).
23. Holmes, J. R. *Practical Waste Management* (New York: John Wiley & Sons, 1983).
24. Flynn, N.W., M. Guttman, J. Hahn, and J.R. Payne. *Trace Chemical Characterization of Pollutants Occurring in the Production of Landfill Gas from the Shoreline Regional Park Sanitary Landfill, Mountain View, California*. Prepared for the Pacific Gas and Electric Co. and the U.S. Department of Energy by Science Applications, Inc. (1981).
25. Zimmerman, R.E., G.R. Lytynyshyn, and M.L. Wilkey. *Landfill Gas Recovery — A Technology Status Report*. NTIS #DE84-001194, ANL/CNSV-TM-12 (August 1983).
26. Wilkey, M.L., R.E. Zimmerman, and H.R. Isaacson. *Methane from Landfills: Preliminary Assessment Workbook*, Argonne National Laboratory Report ANL/CNSU-31 (1982).
27. Diaz, L.F., G.M. Savage, and C.G. Golueke. *Resource Recovery from Municipal Solid Wastes* (Boca Raton, FL: CRC Press, 1982).
28. Haxo, H.E. et al. ''Lining of Waste Impoundment and Disposal Facilities,'' SW-870, U.S. EPA, Cincinnati, OH (1983).

29. Waksman, S.A., *Soil Microbiology* (New York: John Wiley & Sons, 1952).
30. Gotaas, H.B. *Composting* (Geneva: World Health Organization, 1956).
31. Fry, L.J. and R. Merrill. *Methane Digesters for Fuel Gas and Fertilizer*. Newletter No. 3. (Santa Cruz, CA: New Alchemy Institute, 1973).
32. Finney, C.D. and R.S. Evans, Jr. "Anaerobic Digestion: The Rate Limiting Process and the Nature of Inhibition," *Science*, 190(4219):1088, 1089 (December 12, 1951).
33. Sullivan, J.L., C.M. Ostrovski, and N. Peters. "Feasibility Analysis Model for Centralized Methane Production from Animal Wastes," Report for the Ontario Ministry of Energy (File No. 1903-02-876-24) Faculty of Engineering Science, The University of Western Ontario, London, Ontario, Canada (1978).
34. Singh, Ram Bux. "Building a Bio-Gas Plant," *Compost Sci.*, 13(2):12–17 (March-April 1972).
35. Fry, L.J. and Merrill, R. *Methane Digester for Fuel, Gas, and Fertilizer*, Newsletter No. 3 (Santa Cruz, CA: New Alchemy Institute-West, 1973).
36. Khandelwal, K.C. "Dome-Shaped Biogas Plant," *Compost Sci.*, 19(2):22–23 (March-April 1978).
37. Holmes, J.R. *Practical Waste Management* (New York: John Wiley & Sons, 1983).
38. Flynn, N.W., M. Guttman, J. Hahn, and J.R. Payne. *Trace Chemical Characterization of Pollutants Occurring in the Production of Landfill Gas from the Shoreline Regional Park Sanitary Landfill, Mountain View, California*, Prepared for the Pacific Gas and Electric Co., and the U.S. Department of Energy by Science Applications, Inc. (1981).
39. Pfeffer, J.T. "Reclamation of Energy from Organic Refuse," Solid Waste Program, EPA Grant No. EPA-P-800776, Department of Civil Engineering, University of Illinois, Urbana (April 1973).
40. Pfeffer, J.T. "Anaerobic Processing of Organic Refuse," Proc. Bioconversion Energy Research Conf., Institute for Man and His Environment, University of Massachusetts, Amherst (1973).
41. Klass, D.L. and S. Ghosh. "Fuel Gas from Organic Wastes," *Chemtechnology*, 3:689–698 (November 1973).
42. Ghosh, S. and D.L. Klass. "Conversion of Urban Refuse to Substitute Natural Gas by the Biogas Process," presented at Fourth Mineral Waste Utilization Symposium, Chicago (1974).
43. "Fuel Gas Production from Solid Waste, 1974 Semi-Annual Progress Report," NSF Contract C-827, Dynatech Report No. 1151 (1974).
44. Gossett, J.M. and P.L. McCarty. "Heat Treatment of Refuse for Increasing Anaerobic Biodegradability," NSF Grant GI-43504, Dept. of Civil Engineering, Stanford University (1974).
45. Kayhanian, M., L. Linderauer, S. Hardy, and G. Tchobanoglous, "Two-Stage Process Combines Anaerobic and Aerobic Phases," *BioCycle,* 32(3):48–53 (March 1981).

# CHAPTER 10

# Integrated Waste Management

## INTRODUCTION

Recently, several government entities throughout the U.S. and other countries began to require that communities reach levels of recycling that vary between 15 and 50% of the quantities of wastes generated. In order to achieve these levels of diversion from the landfill, it is necessary to plan and implement an overall waste management program such that the various elements of the program are compatible with one another. Compatibility of program elements has led to the widespread use of the term "integrated" waste management. In solid waste management nomenclature, the term "integrated" would be reserved for systems, schemes, operations, or elements in which the constituent units can be designed or arranged such that one meshes with another to achieve a common overall objective. Moreover, the forming or blending must be such that it establishes a community- or region-wide hierarchy for integrated waste management. In many instances, integration has been misinterpreted and has primarily focused on the treatment of the solid waste, i.e., source reduction, recycling, and composting. However, a fully integrated waste management program encompasses much more than treatment. A successful integrated program must consider storage, collection, transport, processing, and final disposition. All of these elements must be developed such that they are compatible with one another. In addition, they must be strategically supported by other relevant programs such as behavioral patterns, economic conditions, public education, public relations, and training.

Integration implies a design such that the activity or operation as a whole functions harmoniously and efficiently — each unit contributing and complementing its neighboring unit. A major portion of this chapter is devoted to explaining the relationship between true integration and beneficial outcome. Approaches are described and examples (especially co-composting) are given of true integration.

## PRINCIPLES

### Definitions and Meanings

Logically, a discussion on integrated waste management — or any other integrated entity or activity — should be begun with some words about the true meaning of "integrated." Unfortunately, "integrated" has become a popular appellation, and consequently, one is almost sure to encounter "integrated" in the labels and titles of proposed waste management undertakings and even in those of some not-so-new undertakings. Upon judging whether the attribute "integrated" is merited, one should have a clear idea of the meaning of the term and its implications. Before continuing the discussion, it should be emphasized that the term "integrated" implies more than a collection of unrelated or independently functioning units under a single roof or management. Of course, such an interpretation is an exaggeration, but it does serve to illustrate the misuse of the term and set the stage for discussing its true meaning and implications.

Of the four versions of the definition of the verb "to integrate," the one that probably is most appropriate to the purposes of this chapter is "to form or blend into a whole." Accordingly, in solid waste management nomenclature, the term "integrated" would be reserved for systems, schemes, operations, etc., in which the constituent units can be and are designed, fitted, or arranged such that one meshes with another to achieve a common overall objective. (As used herein, "unit" may refer to a piece of equipment, to a unit operation, to an activity, or to any other discrete entity.) Moreover, the forming or blending must be such that overlapping is at a minimum and each unit functions at full capacity without interfering with other member units. On the contrary, each unit must function in such a manner that the functioning of the other constituent units is facilitated.

From the preceding definitions and discussion, it is apparent that "integrated" is neither synonymous with "complex," nor does it imply any particular set of dimensions. Thus, an integrated solid waste management operation or system possibly could involve only a few unit operations. An example could be one involving only storage, collection, and disposal under a single management (i.e., individual or corporation). However, in modern usage, the trend is to expand the application of the term to include the institutional as well as the managerial and the technological aspects of waste management.

### Distinction Between Integration and Strategy

The discussion of definitions and meanings is closed with a few words on the distinction between integration and strategy. With respect to the distinction, integration refers to the doing, whereas strategy implies how to go about making the integration. Thus, strategy may determine the type of a particular approach or it may be a pattern for integration.

### Relationship Between Integration and Beneficial Effects

Our discussion of the principles of integration is designed to show the intrinsic relationship between integration and the benefits associated with it. Integration accomplishes its beneficial effects by reducing complexity to simplicity through unification of order from disorder and uniformity from diversity. Simplicity and order leave little room for overlapping or needless duplication of functions. By doing so, it promotes efficiency. Raising the efficiency level, in turn, conduces a lowering of requirements. Regarding the latter, the following especially comes to mind: equipment capacity, power consumption, and land usage. Further meanings and additional implications of these beneficial effects become more comprehensible and, therefore, apparent as our discussion progresses.

With respect to benefits, it should be emphasized that the present situation in solid waste management is such that all of the benefits associated with integration are essential to the success of practically all solid waste management undertakings, especially those that involve resource recovery. In conclusion, the scarcity of available land and the existence of severe disposal siting problems have become so acute as to make resource recovery imperative. Therefore, it follows that a high level of integration must be a feature of current and most future solid waste management undertakings.

## APPROACHES TO INTEGRATION

Once the principles and benefits of integration have been defined and explored, the next step in this Chapter is the more difficult one of attempting to present a strategy for accomplishing it.

### Potential for Integration and Examples

We begin this discussion by tracing the steps in which integration may be of use in a solid waste management strategy. Typically, the evolution of a solid waste management strategy begins with the development of a rationale and proceeds through a series of steps and activities until the time for implementation arises. For purposes of this Chapter, we regard integration as entering the developmental flow at the final or implementation stage.

At the implementation stage, integration can enter the process at a number of levels. Listed in descending order of complexity, these levels are (1) the management or departmental level (storage, collection, disposal), (2) facility level (interfacilities and intrafacility); and (3) process level (inter- and intra-levels). Another set of levels includes the following: (1) plant (within the plant or between plants); (2) facility; (3) technical; and (4) institutional level.

Integration at the management or departmental level would apply to planning, policy making, decision making, and to assignment of responsibilities. An example of integration at the institutional level would be the fitting of solid waste management into an overall environmental improvement plan, or into a resource conservation plan — whether the resource be water, land, or air, or a plan embracing all three resources.

With respect to integration at the interprocessing level, integration could be in large part the arrangement of the sequence of unit processes. The unit processes can be arranged such that the outputs or rejects from one can serve as inputs to another. Furthermore, the inputs must be in such a form so that the design can take advantage of the efficiency of the unit. For example, it has been found that preceding the size reduction unit with a trommel screen enhances the efficiency of the shredding operation.[1] Another example would be the inclusion of a magnetic belt before refuse-derived fuel is introduced into a densifier. In this case the magnetic belt would function as a unit to reclaim magnetic metals and as a protection for the densifier.

An example of integration at the plant or facility level would be the use of treated effluent from a wastewater treatment plant for the quenching of ash from an incinerator or scrubbing of fly ash from stack emissions.[2] The treated wastewater can also be used to supply the water needed in hydraulic separation (e.g., flotation) and/or pulping in resource recovery from the solid waste stream. Energy generated or recovered in the processing of solid wastes (e.g., incineration, biogasification) can be used in the wastewater treatment facility. Many benefits can be derived from the integration of two or more facilities. Some of these benefits include conservation of resources, siting, sharing of maintenance facilities, use of same labor force, and others. Unfortunately, at the present time, the benefits seem to be overshadowed by very strong institutional obstacles. The authors have proposed and evaluated integrated energy-agrowaste processes such as that shown in Figure 10.1.[10]

Integration at the plant or facility level is one of the types of integration that receives the most attention. This is also one that lends itself to representation through "conceptual" designs.

Perhaps one of the most interesting examples of integration is co-composting. Because of the intensity of the many problems that are besetting solid waste management, the search for solutions is becoming increasingly desperate. Among the solutions being advanced, co-composting ranks with the most promising. Because of that fact, considerable attention is directed to the practice in this chapter.

## CO-COMPOSTING

### Historical Development

The concept of co-composting was a natural accomplishment to that of composting municipal refuse. At this point it should be pointed out that in this chapter the term "municipal solid waste" is taken in its popular rather than in its restrictive generic sense. In its popular sense, it usually excludes waste solids resulting from water and wastewater treatment (of "hydraulic" origin). In other words, "municipal solid wastes" applies to those wastes usually catalogued in the average solid waste management textbook as "refuse," "garbage," etc.

The course of the concept of co-composting can be roughly divided into two periods, namely, the "old" (pre-1975) and the "modern" (post-1975). However, as the succeeding paragraphs show, the only real difference between the two periods was one of emphasis.

In the "old" period, the emphasis was on the refuse fraction of MSW, and the sewage sludge was regarded principally as a source of nitrogen and

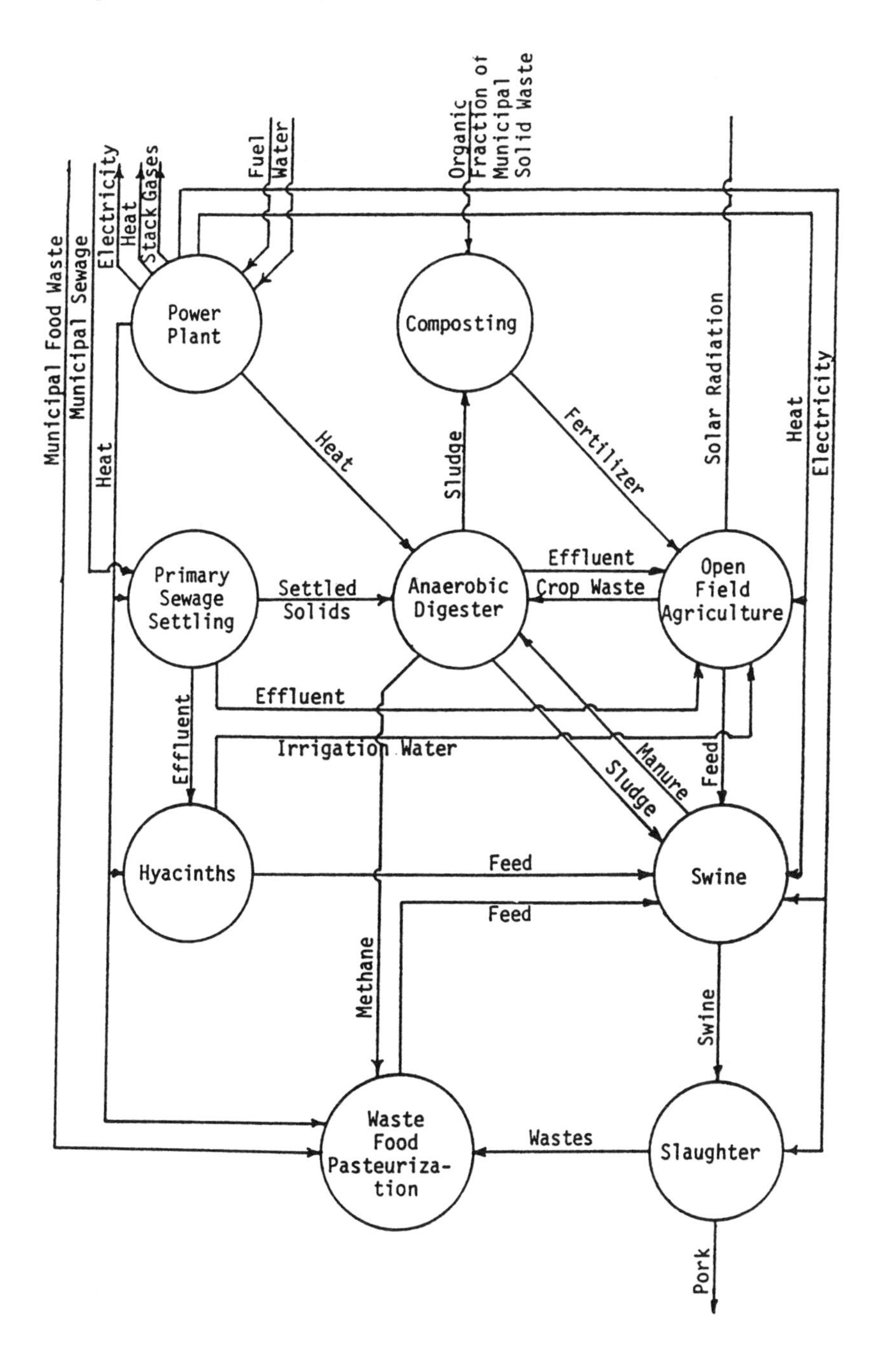


**Figure 10.1.** Diagrammatic sketch showing mass and energy interrelationships of a sewage–hyacinth–digester–swine–composting–open field complex.

phosphorous, and to a lesser extent of potassium for enriching the refuse and thus promoting the compost process. Thus, it might be said that co-composting was a means to an end, rather than an end in itself.

A major reason for the limited role of co-composting was the high moisture content of the sludge (settleable wastewater solids) generated and collected as a part of the treatment of wastewater. At the time, the amounts and impacts of this sludge apparently were sufficiently modest as not to be a cause of concern. Moreover, the technology of sludge dewatering was in its "infancy;" hence, sludge was subjected to very little dewatering before being treated, or in this case, before being added to the refuse. Consequently, very little sludge could be added to the refuse without raising its moisture content to inhibitory levels with respect to the composting process itself. More importantly, at least in terms of the environment and public tolerance, the excess moisture usually led to the development of intensely objectionable nuisances and the production of a noxious leachate.

The second half of the 1970s, i.e., the "modern" period, witnessed a drastic shift in the emphasis of composting, i.e., from the refuse fraction to the sludge fraction. The shift was largely a response to a desperate need for a means of treating sewage sludge other than by way of incineration or landfill. Composting seemed to meet the need, especially since sludge treatment had been considerably advanced, particularly with respect to dewatering. In the new setting, sludge was dewatered to a solids content at which composting became feasible provided that a bulking agent be used. Although not particularly troublesome at first, this stipulation in recent years became a financial handicap. With the worsening of the financial viability came a renewed interest in the possibility of composting refuse. It is not surprising that the possibility of bypassing the requirement for bulking by using the organic fraction of refuse as a bulking agent gained attention, thus, the current interest in co-composting.

## Methods

Except for the provision for the addition of sewage sludge or refuse, as the case may be, the steps in co-composting, regardless of compost method used, are the same as for composting any other organic residue. In co-composting, the sequence is as follows: (1) collect and process the yard waste (if yard waste is to be included); (2) segregate and prepare the organic fraction of refuse; (3) mix the materials with sludge cake (or vice versa); (4) compost the mixture; and (5) prepare the composted product for storage, use, or sale.

The sections that follow in this Chapter on co-composting deal with certain key requirements, five commonly encountered problems, and suggested solutions to the problems. To illustrate these problems and solutions, an abbreviated case history is given of a compost plant.

## Problems and Solutions

The five problems to receive attention are (1) adequate mixing of the sludge and the refuse; (2) insufficient primary (front-end) processing; (3) inadequate management of moisture content (too dry or inhibitively wet); (4) tendency of the composting mass to arch or bridge over when mechanical mixing devices are used; and (5) lack of understanding of the composting process on the part of the operators.

### *Mixing*

Mixing refuse and sludge involves two vitally important considerations, namely, (1) the proper ratio of sludge to other wastes; and (2) the procedure followed for mixing the wastes to form a homogeneous mass. Many factors enter into a determination of a proper or suitable ratio. Among them might be whether the disposal of refuse supersedes that of sewage sludge, or vice versa. If refuse disposal is uppermost, then the ratio of refuse to sludge would be as high as consonant with successful composting. If sludge is the cause of greatest concern, then obviously the reverse would be true. However, if the disposal of both is of equal concern, then the ratio should be such that the relative generation rates of both wastes would be matched. Fortunately, all three of the general ratios can be used with equal success by suitably adjusting the respective moisture contents of the constituents, and taking care that the C:N of the mixture be at a suitable level.

A thorough mixing of refuse with sludge cake (or vice versa) is essential to the success of a co-compost operation. Unfortunately, it also is one of the more difficult tasks to accomplish. Ideally, the mixture should be completely homogeneous and yet be porous. Moreover, this porosity should be maintained throughout the duration of the composting process. Although the addition of yard waste contributes to the maintenance of porosity, a mixture solely of refuse and sludge has an innate tendency to settle (i.e., compact) whether it be in a windrow or in an enclosed reactor. The tendency is not surprising in view of the fact that refuse in the U.S. has a large concentration of paper and sludge cake typically is on the order of 70 to 80% water.

The viscous nature of sludge cake (comparable to wet clay) renders homogenizing it with refuse a particularly difficult and onerous undertaking. A major difficulty is the tendency of the two wastes to agglomerate into large clumps and globose masses. Although the mixing problem can be met with the use of a machine designed by the authors,[7] the better approach would be to use a dried sludge.

### *Insufficient Primary Processing*

Primary processing of refuse for both composting and co-composting is step 2 in the sequence, namely, segregate and prepare the refuse. The importance of adequate primary processing lies more in the eventual quality of the compost product than in the course of the biological process that constitutes composting. If the feedstock contains substances that are nonbiodegradable (hereinafter referred to as "contaminants"), they would have little or no effect on the biological processes. This, however, would not be the case if the substances are biologically toxic.

At the least, the nontoxic contaminants generally are of such a nature and appearance as to discourage customer acceptance. Additionally, they may lower the agricultural utility of the product by interfering with the performance of crop care and harvesting equipment. Their persistence may even be such as to render harvested crops unsuitable or harmful to the animals for whom they were intended. Film and other forms of plastics as well as glass shards, however minute in size, are key offenders in this respect.

Leading examples of contaminants are metal and plastic beverage containers, bottle caps, all forms of plastics, and glass shards, regardless of origin. It is far easier to remove such materials from the raw material introduced into the composting process than to attempt to remove them from the product at the end of the process.

Deterioration of product appearance and utility is but one of the undesirable effects. Such materials contribute to the volume of the materials to be passed through the compost units, thus reducing capacity and increasing cost per unit of material treated. Moreover, the material must be handled along with the material to be composted, thus magnifying handling charges.

### Extremes of Moisture Content

In general, compost operators have become well aware of the problems that are generated by an excessively high moisture content (lack of porosity, promotion of anaerobiosis, etc.) and, accordingly, make every effort to avoid it. Unfortunately, a new trend seems to be developing, namely, failure to provide an adequate supply of moisture. A leading contributor to the trend is the fact that all too often dryness is equated with maturity of product. It should be remembered that all biological activity ceases after the moisture content descends to 8 or 12%. Occasionally, the product may heat up at the low levels. Any such heating is strictly chemical or physical in origin. As far as composting is concerned, a safe and yet adequate range of moisture content for refuse and similar wastes is on the order of 40 to 55%.

### Arching or Bridging During Composting

Arching or bridging are mechanical problems often noted in mechanized composting during mixing or transfer of the composting material. Instead of dropping into place in the mixing or transfer device, the material arches over the device such as to produce a tunneling, leaving the material unmixed or not transferred. This phenomenon is partly due to equipment design and is aggravated by the physical nature of the composting product.

### Inadequate Understanding of the Compost Process

An unfortunate and persistent problem is the lack of understanding of the compost process on the part of many operators. This problem is especially pronounced with respect to both the preparation of the substrate (particularly mixed municipal solid waste) as well as the biological principles of composting. Although a thorough understanding of the mechanical aspects and an awareness of conventional operational parameters may be had by the operator, they are of little use in deciding what are the remedial measures to be taken when unexpected developments occur in the composting process. The lack of understanding results in inefficient use of equipment and may lead to the avoidable and therefore unnecessary development of nuisances and undesirable deterioration of the surrounding air, water, and soil environments, as well as the production of an inferior product.

### Case History

The five problems described in the preceding sections were observed by the authors to be in full force in the operation of a composting plant. The plant was placed in operation as part of an overall recycling effort. In fact, the difficulties might be said to be ultimately due to the plant being a part of the recycling effort.

In designing the refuse collection aspects of the recycling program with respect to the role of the citizenry, the segregation to be made by the citizen in discarding his or her wastes was only into two fractions (containers), namely inorganics and organics, i.e., into rubbish and garbage fractions. As one would expect, the outcome is that the rubbish containers receive only the easily distinguishable and classifiable items, plus a few organics. All remaining material biodegradable and nonbiodegradable is discarded into the second container. Unfortunately, little or no provision has been made at the compost facility for further segregation and processing. As a consequence, a heterogeneous mixture of compostable and noncompostable material is discharged into the compost unit. No provision is made for removing the contaminants from the finished product, which consequently is visually unsatisfactory.

The lack of understanding of the compost process is manifested in the unsatisfactory performance of the operation and the continued insistence on the part of the operator that all is well. Unless these conditions are recognized and suitable remedial measures are taken, it is almost inevitable that undesirable conditions will soon reach a level at which they can no longer be ignored.

## INTEGRATED RESOURCE RECOVERY SYSTEMS

### Introduction

This chapter is closed with a brief mention of two systems selected by the authors from several designs for integrated systems proposed by them in the past. The combination of subsystems used in accomplishing integration in the system is based upon actual trials and experience. Only the highlights of the systems are presented at this time; they are described in some detail in References 8 and 9.

## Potential of System I

The system was designed to recover a variety of products and materials from urban solid waste. It involves the systematic removal of economically viable products in a sequence that optimizes purity of the end product and economical recovery, and maintains environmental impacts to a minimum. In the system, the various fractions (i.e., recyclable materials) are isolated along with a highly organic fraction suitable for biological treatment. The organic material can be composted or digested anaerobically. The composting process can be carried out with or without the addition of sewage sludge.

The integrated system provides for direct recovery of about 60% of the incoming solid waste stream in the form of recyclable secondary materials, along with the capability of increasing the recovery to 80% through the addition of processing subsystems.

### *Design*

The system is of modular construction. This type of construction allows for several different or combined processing subsystems. The system described herein basically consists of a dry plant and a wet plant.

The dry plant provides primary size reduction, air classification, and screening. The "heavies" and "lights" from the air classifier are processed separately in the heavies and lights recovery systems. The heavies are subjected to ferrous removal. Following screening to remove fine organic and inorganic matter, the lights may be used directly as a fuel; or as a feedstock for recovery for wastepaper pulp or other fibers; or, after further processing, as an exceptionally clean, homogeneous fuel.

The light and the heavy organic fractions may also be used to provide the feedstock for a composting system. After removal of the greater part of the inorganic materials, the two fractions are combined and mixed with sewage sludge and are composted. One of the authors was also able to mix a portion of the organic fraction from the refuse with raw sludge and digest the mixture anaerobically. The solid residue from the digestion process would serve as an excellent feedstock for composting.

The wet processing facility consists of a pulper, a plastic removal screen, pressurized screen, and a bank of centrifugal cleaners. A conventional wastewater treatment facility treats wastewater generated in the process. Strength properties of pulps recovered from residential and commercial refuse compares favorably with fiber from 100% deinked newspapers and spruce groundwood. Although both burst and tear properties tend to be greater than those for the deinked newspapers and spruce groundwood, the breaking-length values of the pulp from residential sources tend to be somewhat lower, whereas that from commercial solid waste is similar. A diagram of the system is given in Figure 10.2.

## Potential of System II

Solid waste facilities that will serve as major elements of waste diversion strategies include MRFs for source-separated recyclables, yard waste processing

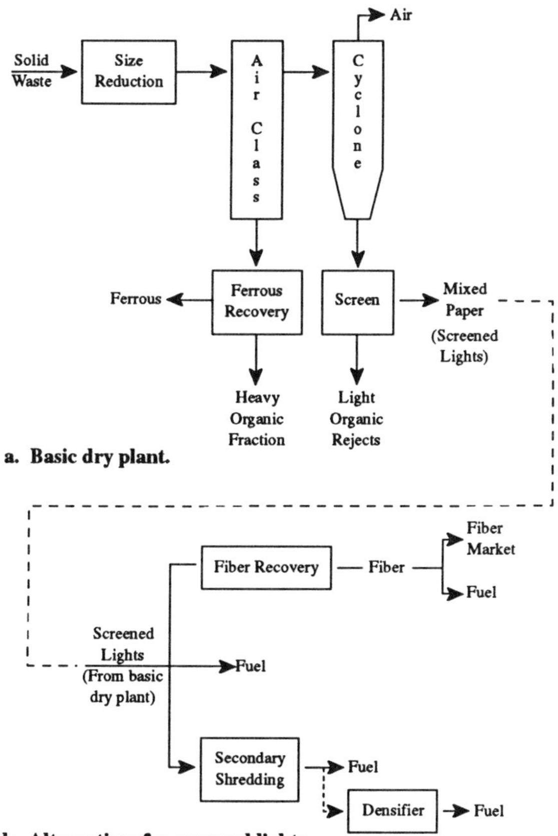


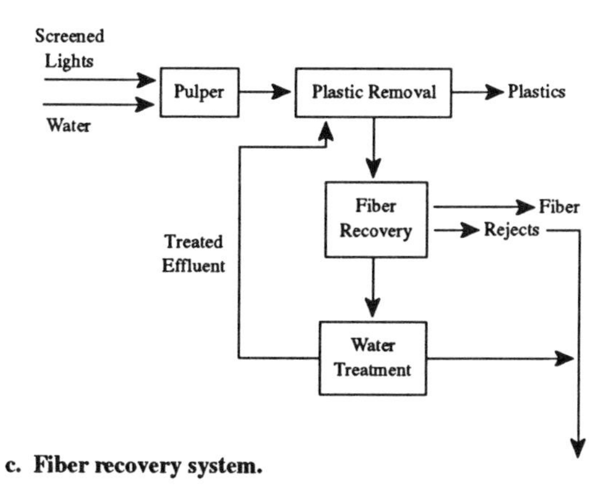


**Figure 10.2.** Integrated resource recovery system.

operations, and mixed MSW processing systems. This section describes the design of MRFs, yard waste processing operations, mixed MSW processing systems, and means of their integration. Examples of integrated waste management strategies are presented, illustrating the design and integration of subsystems.

The seriousness of the technological challenges represented by diversion goals of 25 to 50% can be gleaned by examining the composition of a typical municipal solid waste stream and the requisite recovery percentages. On the order of 70% recovery likely is necessary to achieve diversion percentages of 50% or greater, as illustrated in Table 10.1. The recovery percentages shown in the table may be composed of separate percentages applied to the collection systems and to the processing systems.

The fact is that the 70% recovery of the components listed in the table is difficult to achieve. Presently, there is little historical support for such levels of recovery. However, to meet a 50% diversion objective, the difficulties must be overcome. The remedies include public education, program promotion, and adequate stable markets and secondary material prices.

In this chapter we will examine the technical means to reach high percentages of waste diversion. In the present case, we assume that markets are available for the baseline recyclables (tin cans, etc.), yard waste (as hog fuel, mulch, or compost), and end products derived from other waste components (e.g., corrugated, mixed paper, food waste, and wood). The assumption of availability of markets for the recovered materials, while a convenience to simplify this discussion, is not meant to minimize the substantial importance of the definition of product specifications and the negotiation of written end product purchase agreements early during the development of the project.

### Design

The approaches to the design of this type of integrated solid waste management processing system can be divided into two, based on the forms of the solid wastes delivered to the facility. The forms of the delivered waste, i.e., the feedstocks, are mixed waste and source-separated waste. The characteristics of the two forms of waste are dramatically different in terms of contamination of individual components (e.g., newspaper, aluminum cans, etc.) by other components such as food waste, dirt, etc. The dramatic difference in levels of contamination necessitates different methods of processing of the two feedstocks.

Source-separated wastes obviously possess low levels of contamination if the program is properly administered. Mixed waste, as the term implies, is a heterogeneous mixture of solid waste components. At this time, source-separated wastes typically consist of approximately three to six components (e.g., tin cans, aluminum cans, glass containers, newspapers, PET, and HDPE), while the constituents of mixed wastes can number in excess of three dozen material categories. Source-separated wastes can also include those of one waste component, such as yard waste or tires.

**TABLE 10.1 Example of Situation of Achieving in Excess of 50% Waste Diversion**

| Components | Percent in MSW | Recovery Percentage | Recovered (%) |
|---|---|---|---|
| Baseline recyclables | | | |
| Tin cans | 3.0 | 70 | 2.1 |
| Aluminum | 0.5 | 70 | 0.4 |
| Glass | 8.0 | 70 | 5.6 |
| News | 12.0 | 70 | 8.4 |
| Total | 23.5 | 70 | 16.5 |
| Yard waste[a] | 15.0 | 70 | 10.5 |
| Other wastes[a] | | | |
| Corrugated | 10.0 | 70 | 7.0 |
| Mixed paper | 20.0 | 70 | 14.0 |
| Food waste | 5.0 | 70 | 3.5 |
| Wood | 2.0 | 70 | 1.4 |
| Total | 37.0 | 70 | 25.9 |
| Totals | 75.5 | 70 | 52.9 |

[a] For example, diverted from landfill disposal via conversion to a compost product.

The subsystems of the integrated waste management system discussed herein are

1. Material processing facilities which process source-separated recyclables (i.e., recyclable metal, glass, plastic containers, and paper) for market.
2. Subsystems which process mixed solid wastes for recovery of recyclable containers and waste-derived compost, secondary paper fiber, or fuel.
3. Subsystems which process yard waste into a hog fuel, compost, or mulch.

Each of the subsystems can be integrated with one or more of the other two subsystems with proper attention given to planning and engineering design. The design goal, of course, is the processing of the raw feedstock into a marketable or usable end product based on a set of specifications. To achieve a substantial percentage of diversion of generated wastes from landfill disposal requires careful consideration of feedstock properties, processing equipment performance, and local markets.

## SOURCE-SEPARATED PROCESSING SYSTEM DESIGN

Process designs for an MRF are shown in Figures 10.3 and 10.4, respectively, for a paper processing line and a container processing line. The process design in each of the illustrations assumes that recyclables arrive at the facility in source-separated, singular form (e.g., tin cans) and in commingled form (i.e., mixtures of several source-separated material categories). The design of each processing line provides for redundancy in receiving, manual sorting, mechanical processing, and efficient removal of contaminants.

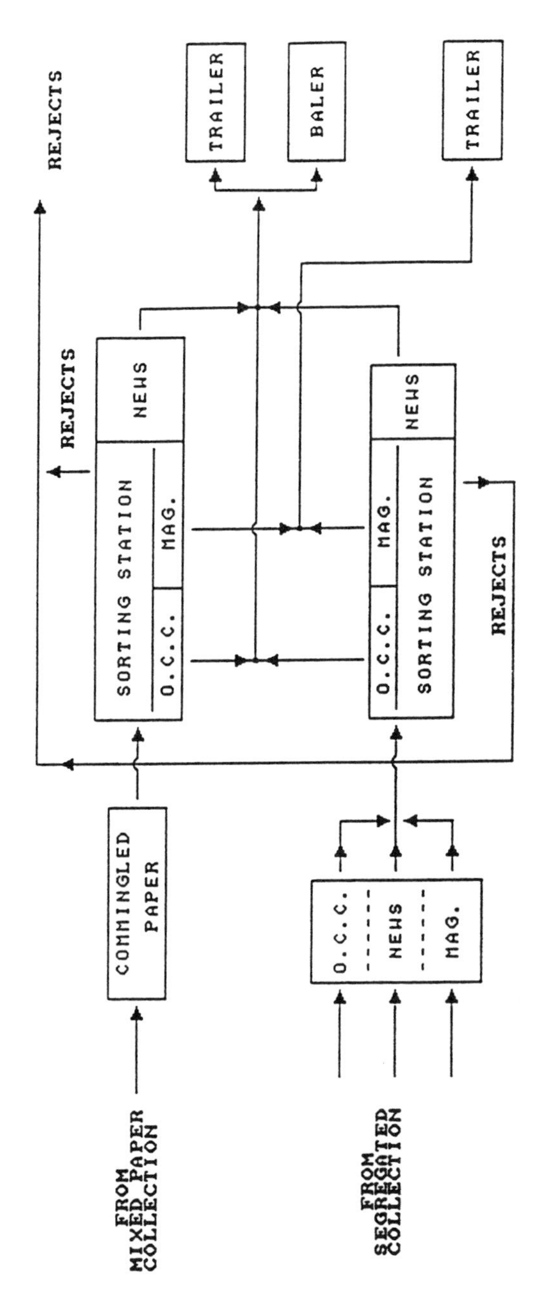


**Figure 10.3.** Source-separated paper processing line.

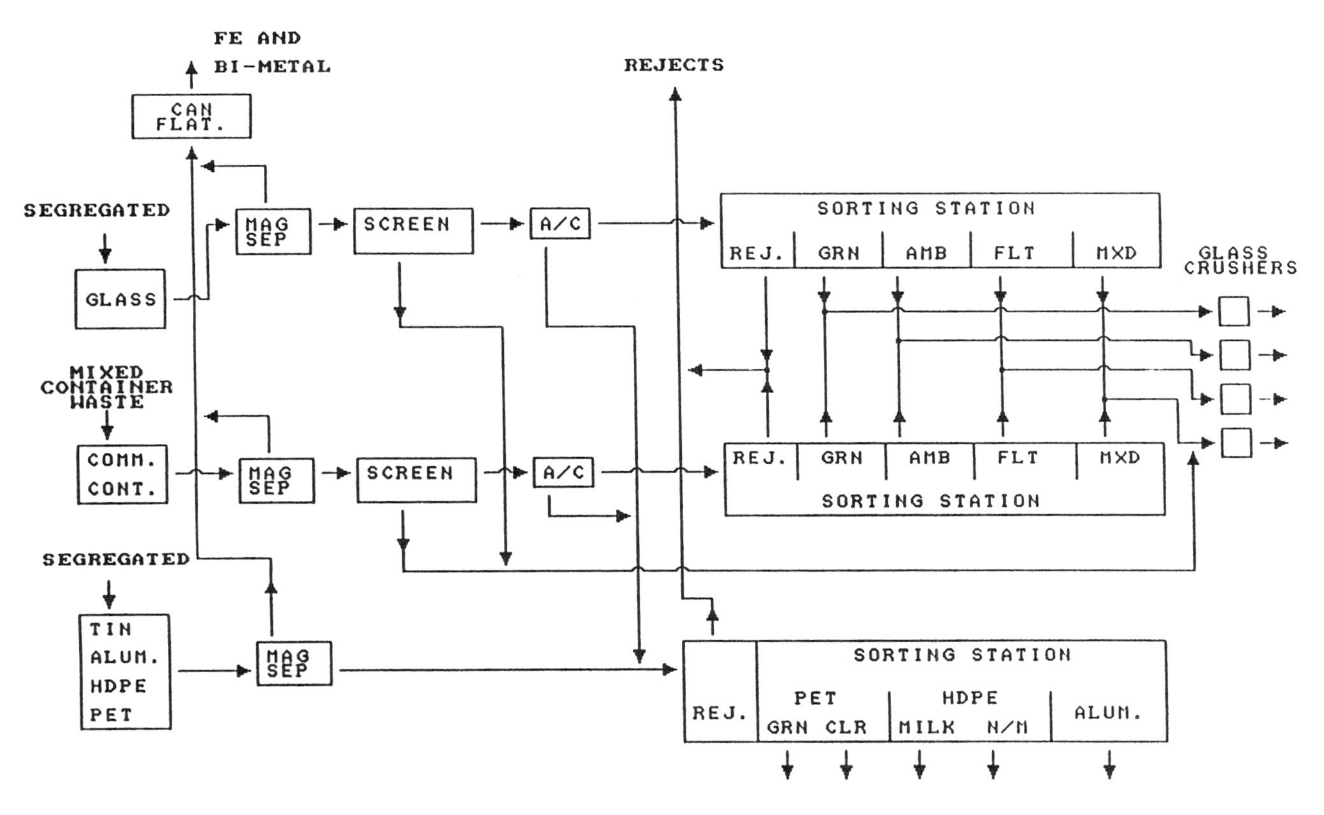


**Figure 10.4.** Source-separated container processing line.

## MIXED MSW PROCESSING SYSTEM DESIGN

As opposed to MRFs processing commingled and segregated material categories, wherein normally 90% or more of the input materials are recovered in the form of marketable end products, facilities processing mixed MSW can recover approximately 10 to 20% of the input in the form of marketable grades of metals, glass, plastics, and paper. Additional material recovery can be achieved by integrating into the facility design additional processing operations to recover refuse-derived fuel (RDF), a compostable feedstock, or secondary paper fiber. These options for integration can increase the total diversion from landfill disposal to within the range of 60 to 85% of the input MSW if markets for the other materials exist.

An example of a facility design configured for the primary purpose of processing and recovering recyclable materials from mixed MSW, including ferrous, HDPE, PET, aluminum, and several grades of paper (corrugated [OCC], newspapers, and high grades), is illustrated in Figure 10.5. The processing system incorporates both mechanical and manual separation processes in order to optimize the recovery of marketable secondary materials. The design recovers approximately 15% of the input mixed waste in the form of marketable grades of recyclables.

Provision is made in the facility design to segregate corrugated and other marketable waste paper grades by wheel loader that arrive in loads of waste composed predominantly of paper materials. When sufficient corrugated or other paper grades are removed on the tipping floor by wheel loader and accumulated, the materials are transported directly to a baler, bypassing the mixed waste processing equipment.

The key unit operations of the process design are trommel screening, magnetic separation, and manual sorting, as shown in Figure 10.5.

For the material recovery scenario, discussed in the preceding paragraph, process residues account for about 85% of the incoming mixed solid waste. Much of the process residues are combustible and biodegradable organic components composed predominantly of cellulosic materials. These process residues require landfill disposal unless processed for energy recovery, converted to a compostable feedstock for subsequent composting, or processed to produce waste-derived secondary paper fiber. For example, if RDF recovery is integrated with materials recovery from mixed MSW, the residue stream could be reduced to 15 to 25% of the input MSW. In the case of recovery of secondary fiber or of a compostable feedstock, the residue stream would be in the range of approximately 45 to 55% and 30 to 45%, respectively.

## YARD AND WOOD WASTE PROCESSING DESIGN

Several alternatives are available to utilization of yard and wood wastes. The alternatives include the production of compost, sawdust, and hog fuel. For either alternative, and assuming relatively uncontaminated waste materials, the

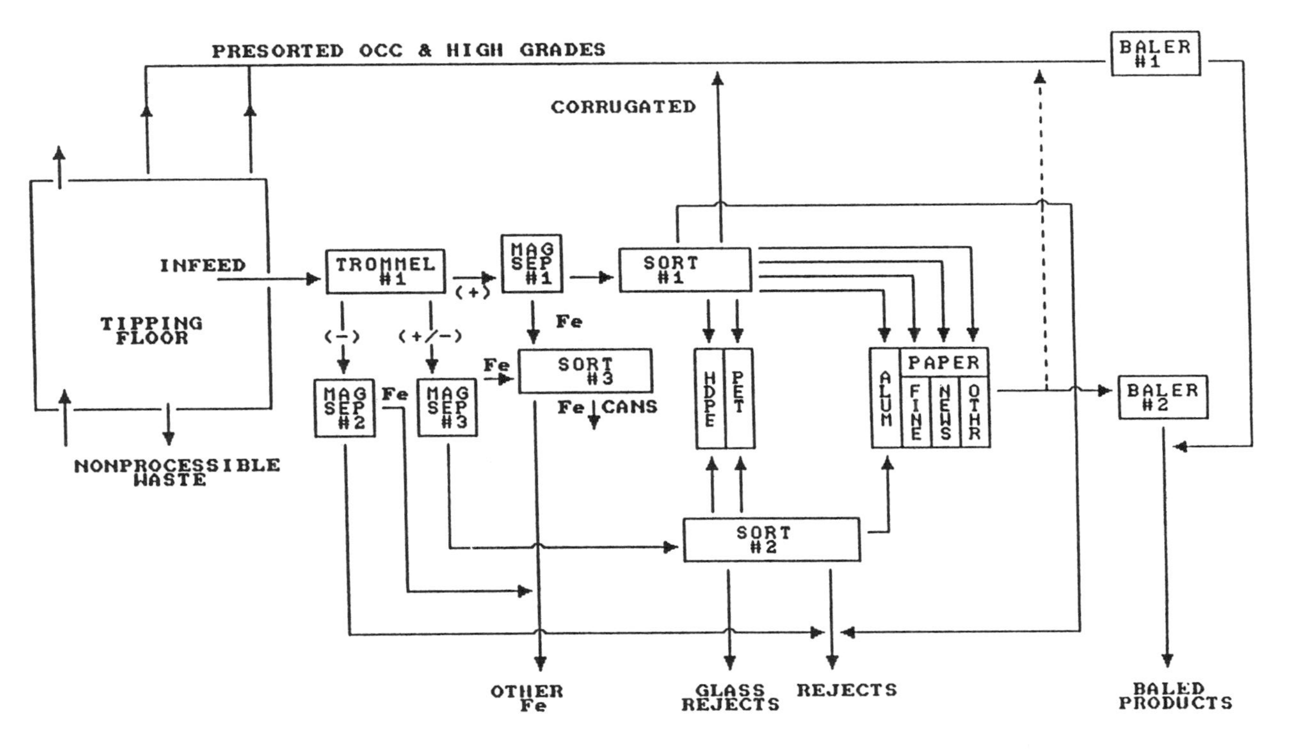


Figure 10.5. Mixed MSW processing system.

necessary unit operations are size reduction, magnetic separation, and screening, as shown in Figure 10.6. Magnetic separation is employed to remove ferrous contamination (e.g., nails, staples, etc.) from the size reduced wood and yard waste product. In the case of the production of a hog fuel, usually a relatively large particle size is desired. The fines either are used as a soil amendment or disposed in a landfill. If the process application is the production of a yard waste compost, a finely sized product is the objective.

The production of hog fuel (solid lines) and of a compostable yard waste product (dashed lines) is illustrated in Figure 10.6.

## CONCLUSIONS

Integration of various processes is a necessity in order to meet high levels of waste diversion.

The processing of source-separated residential recyclables and of source-separated yard wastes, using the aforementioned subsystems, typically would result in the diversion of 20 to 30% of generated solid wastes from landfill disposal. If source-separated commercial waste recycling and processing is added to the overall program and properly administered, an additional 5 to 10% diversion typically is achievable. The total diversion combining these programs is in the range of 25 to 40%.

If mixed waste processing for recovery of the remaining non-source-separated metal, glass, and plastic containers and for recovery of compost is incorporated with the subsystems described in the preceding paragraph, the overall diversion level can be increased to the range of 55 to 75%.

The implications of integrated solid waste management systems on landfill disposal capacity extend beyond the mere reduction in *quantities* requiring landfill disposal. The characteristics of process residues (i.e., waste by-products) from integrated solid waste management systems having substantial diversion percentages will be significantly different from that of contemporary MSW. The residues will typically be dense, inorganic materials inasmuch as the majority of the lighter organic materials (e.g., paper and yard waste) have been removed via processing. Among other considerations, landfill volume requirements will have to be calculated based on the properties of the process residues and not on those typical today of MSW.

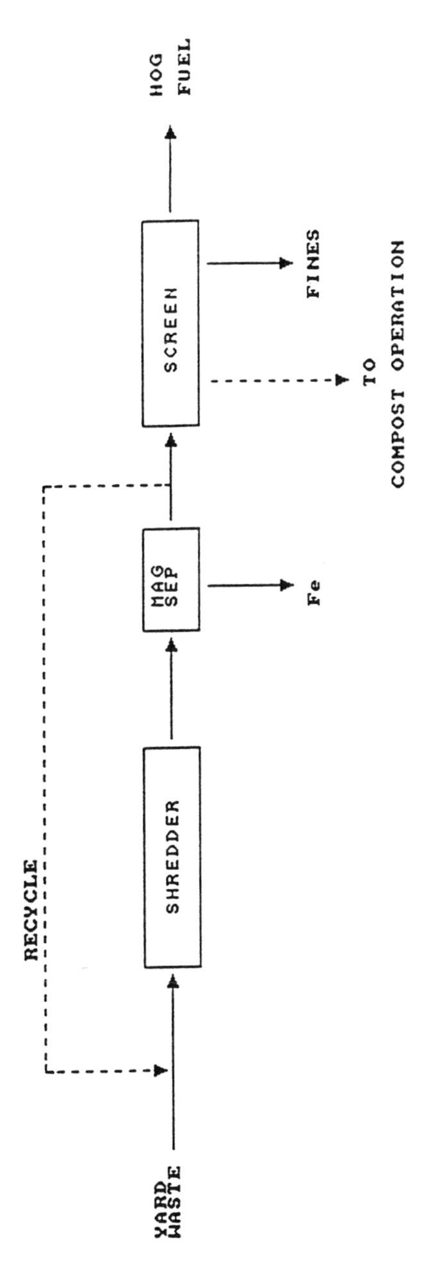


**Figure 10.6.** Yard waste processing.

## REFERENCES

1. "Prediction of Impact of Screening on Refuse Derived Fuel Quality," Report prepared by CalRecovery Systems, Inc., and Midwest Research Institute for the Electric Power Research Institute, EPRI FP-1249 (1979), pp. 6.1–6.8.
2. Diaz, L.F., G.M. Savage, and C.G. Golueke. *Resource Recovery from Municipal Solid Wastes,* Vol. II (Boca Raton, FL: CRC Press, 1982), pp. 27–38.
3. Goldstein, N. "Composting Picks Up Steam," *BioCycle*, 28(9):27–30 (October 1987).
4. "Feasibility Evaluation of Municipal Solid Waste Composting for Santa Cruz County, California," Report prepared by CalRecovery Systems, Inc., for the California Waste Management Board (December 1983).
5. Golueke, C.G. *Biological Reclamation of Solid Wastes* (Emmaus, PA: J.G. Press, 1977).
6. Diaz, L.F., G.M. Savage, and C.G. Golueke. *Resource Recovery from Municipal Solid Wastes,* Vol. I (Boca Raton, FL: CRC Press, 1982).
7. Golueke, C.G., D. Lafrenz, B. Chaser, and L.F. Diaz. "Composting Combined Refuse and Sewage Sludge," *Compost Sci./Land Util.*, 21:42 (1980).
8. Diaz, L.F., G.M. Savage, and C.G. Golueke. "An Integrated Resource Recovery System," *BioCycle*, 28(10):47–52 (November/December 1987).
9. Savage, G.M. "Design of Integrated Solid Waste Management Systems," presented at the Thirteenth Annu. Madison Waste Conf. on Municipal and Industrial Waste, University of Wisconsin, Madison (September 1990).
10. Diaz, L.F. and C.G. Golueke. *Input-Ouput Analysis of Various Elements of an Energy-Agro-Waste Complex*, ORNL Report No. TM-79099 (November 1979).

# Index

## N



## O

## P



## Q



## R

## T



## V



## W



## Y

## Z